Duration, Temporality, Self

Duration, Temporality, Self

Prospects for the Future of Bergsonism

Elena Fell

PETER LANG

Oxford · Bern · Berlin · Bruxelles · Frankfurt am Main · New York · Wien

Bibliographic information published by Die Deutsche Nationalbibliothek
Die Deutsche Nationalbibliothek lists this publication in the Deutsche Nationalbibliografie;
detailed bibliographic data is available on the Internet at http://dnb.d-nb.de.

A catalogue record for this book is available from the British Library.

Library of Congress Cataloging-in-Publication Data:

Fell, Elena, 1975-
 Duration, temporality, self : prospects for the future of Bergsonism /
Elena Fell.
    p. cm.
Includes bibliographical references (p.   ) and index.
ISBN 978-3-0343-0883-0 (alk. paper)
1. Bergson, Henri, 1859-1941. 2. Time. I. Title.
B2430.B43F44 2012
115.092--dc23
                          2012027896

Cover image and front cover design by Dawn Rodd
http://www.flickr.com/photos/crystal_singer/

ISBN 978-3-0343-0883-0

Peter Lang AG, International Academic Publishers, Bern 2012
Hochfeldstrasse 32, CH-3012 Bern, Switzerland
info@peterlang.com, www.peterlang.com, www.peterlang.net

Printed in Germany

Time is real.
— *Henri Bergson*

Contents

viii

Tables and Diagrams

Acknowledgments

First and foremost, I am deeply indebted to Professor Paul Crowther, Dr Andros Loizou and Dr Peter Lucas. I have benefited from their guidance, their talents and their expertise during the years of my research, and their influence on me as a scholar has been truly profound.

Thank you to Dr Johan Siebers and other UCLan colleagues for their friendly support and advice, always readily given. I am very grateful to Rosa Bettess for commenting on the final version of the manuscript and to Anne McCann for proofreading and, of course, this project could not have been completed without the unfailing support of my husband Anthony, who believes in me and always stands by me. Also, I must not forget my son Andrew, whose arrival gave me a lot of inspiration.

A debt is owed to my parents. My father Vladimir Mirenkov, a professional philosopher, introduced me to philosophy from a very early age. My mother Irina Mirenkova together with my aunt Tamara Orlovskaya has provided illustrations for this book. Also, many thanks to Dawn Rodd for designing the cover and to Hannah Godfrey and Holly Catling for helping me to prepare the manuscript for publication.

I should also thank my tutors from St Petersburg State University, especially the Chair of Ontology and the Chair of Contemporary Western Philosophy, who encouraged my interest in the problem of time, and introduced me to the intriguing philosophy of Bergson.

I. V. Mirenkova and T. V. Orlovskaya, *Waiting for Sugar to Dissolve* (2012)

Introduction

Although Bergson explores a wide range of philosophical problems, one could characterize his philosophy as a philosophy of time, where time is not addressed as an abstract notion but is examined as part of real processes, as real embodied time. Discussing the nature of those processes, Bergson inevitably looks at many other issues, and as a result, his reader is faced with a complex discourse where discussions on time are entwined with discussions of memory, matter, intuition, images etc. Some of Bergson's commentators have attempted to interpret his entire contribution to philosophy, while others have addressed particular issues, but undoubtedly all components of his philosophy form part of the whole and cannot be fully comprehended in isolation from other components.

Amongst those who addressed Bergson's philosophy as a whole, both Kolakowski and I. W. Alexander offer a useful and concise overview. More thorough accounts have been made, for example, by Cunningham, Lacey, F. C. T. Moore and Mullarkey. Cunningham embarks on the work of interpretation by arranging Bergson's theory into several topics, dealing separately with intuition, intelligence, duration and finalism. Organizing Bergson's philosophy is a necessary step towards the better understanding of it, but can only be accepted provisionally, because as Čapek observed about elements of Bergson's theory, '[I]t is almost childish to number each individual feature separately, since all of them are complementary and inseparable aspects of one single, though very complex, dynamic reality'.[1] F. C. T. Moore's enthusiastic discourse offers clarifications of many difficult Bergsonian terms and employs examples taken from elsewhere to effectively illuminate and defend Bergson's position. Lacey's study goes further than a mere exposition and clarification: he approaches Bergson from the analytical standpoint and does not refrain from raising difficult questions. In

---

[1]   Milič Čapek, *Bergson and Modern Physics* (Dordrecht: D. Reidel, 1971), 91.

particular, he queries Bergson's concept of pure change, suspicious of his assertion that movement does not require a moving thing – a theme that is important for the Bergsonian interplay of space and time.

Altogether, these authors provide much needed explanations of Bergson's key arguments by systematizing Bergson and elucidating links between parts of Bergson's philosophy. But this is not enough. Deeper analyses of Bergson reveal the need to move beyond what he explicitly states into the realm of principles which are embedded in his work, and which follow from his arguments without, however, being referred to directly. Mullarkey aims at addressing the entire philosophy of Bergson whilst taking this next step. In particular, he treats Bergson's philosophy as dynamic in itself and even refers to it as 'philosophies' of time,[2] rather than merely one philosophy, thereby offering a view that can accommodate certain inconsistencies in Bergson.

My contribution to Bergsonian studies will consist in extracting Bergson's theory of time from his three main texts, *Time and Free Will* (*TFW*), *Matter and Memory* (*MM*) and *Creative Evolution* (*CE*), with references to his other works, *The Creative Mind* (*CM*), *Duration and Simultaneity* (*DS*), *Mind-Energy* (*ME*), *The Two Sources of Morality and Religion* (*TSMR*), *An Introduction to Metaphysics* (*Introduction*). This extraction, as well as offering a concise exposition of this theory, also reveals its incomplete and fragmentary nature, and the remainder of this study consists in an attempt to fill in the gaps and respond to questions which arise along the way. At that stage the debt is owed to those commentators who focus on specific Bergsonian issues. For example, my analysis and further development of heterogeneity was inspired by Čapek,[3] and the discussion of discontinuity would not be complete without references to Bachelard.

It is possible to read Bergson in different ways. One can dismiss his philosophy as Russell does[4] for his refutation of rationality and space; one

2   John Mullarkey, *Bergson and Philosophy* (Edinburgh: Edinburgh University Press, 1999), 2.
3   Čapek, *Bergson and Modern Physics*, 83–185.
4   Bertrand Russell, *The Philosophy of Bergson* (Cambridge: Bowes and Bowes, 1914).

can expect Bergsonism to be a complete and finished theory which should be able to resolve all sorts of philosophical questions; or one can take on board the Bergsonian idea that to exist means to change and, whilst analysing what Bergson explicitly said, allow his philosophy to evolve by working out what he would have said, and what else can be said. I take the latter approach, and the main aim of this project is to indicate a possible way in which the theory of duration can develop further.

I find the biggest attraction of Bergson is in his attempt to grasp the nature of time and show a way of treating time as metaphysical reality, overcoming difficulties humbly admitted to by St Augustine.[5] But Bergson's theory of duration is not a completed, finalized theory. Firstly, it is not put forward in a systematic way and needs to be extrapolated from his more general discourse; secondly, it contains inconsistencies and gaps; and thirdly, it does not address some obvious issues. Moreover, some of Bergson's claims are too strong and need to be examined carefully.

In the expository chapters (Chapters 2, 3, and 4), I examine Bergson's theory of time, which can be called a theory of duration, from his major texts, *Time and Free Will*, *Matter and Memory* and *Creative Evolution*. In each book Bergson introduces duration anew, as if disregarding claims made in previous texts; thus each time duration is given a different, sometimes seemingly opposite, meaning. However, where an unsympathetic critic would see inconsistencies, I see phases of conceptual development of the idea of duration. This being said, Bergson's phases are not linked in a satisfactory way. A key strategy of this study, therefore, will be to fill gaps, raise further questions, and develop new arguments.

I move from duration as a psychological process in *Time and Free Will* to duration as the universal movement in *Creative Evolution*, via the intermediate proposal in *Matter and Memory* that duration is a general principle of being. Duration in Bergson turns out to be an all-embracing concept, itself equivalent to the idea of being. Indeed, can we find in Bergson anything which is not duration? Spatial objects, one may suggest,

5   St Augustine, *Confessions*, transl. R. S. Pine-Coffin (Harmondsworth: Penguin, 1961), Book XI, 264.

reading *Time and Free Will*. But as readers of *Matter and Memory*, we have to accept duration of matter and duration within matter. In *Creative Evolution* everything non-durational is reduced to an illusion, and from this position we can equate 'duration' and 'being'. The main aspect in which the idea of duration differs from the idea of being is that duration already entails a characteristic of being as moving which, according to Bergson, is its necessary feature. What the term 'duration' achieves is to weld motion onto being and demonstrate that being cannot be regarded in any other way than as being in motion, the being that has history. Also, it emphasizes the omnipresence of motion and change, so that even in those cases when we struggle to find and define substance, such as in music or thought, we still find change and motion.

I take on board Bergson's idea that duration is heterogeneous. The idea of heterogeneity emerges when Bergson analyses psychological continuity. Elements of such continuity (emotions, sensations) are not clear-cut, even though we commonly distinguish one emotion or sensation from another. This division, I agree with Bergson, is artificial and done for convenience, as in reality one state of consciousness flows into another and ultimately there is just the unity of the conscious process corresponding to the life of a concrete person.

But this idea of heterogeneity entails a paradox. Although its elements are inseparable, they are different and diverse, so on the one hand Bergson wants us to accept that we cannot individualize them as if they were autonomous units, but on the other hand he does not allow them to be merged into a homogenized stream. In Chapter 5 I attempt to resolve this paradox by claiming that the identity of elements within duration, not given ostensibly, is nevertheless manifested through their unique effects on the world. Bergson says very little about the structure of heterogeneity, and later in Chapter 5 I analyse its composition on a general metaphysical level.

In Chapter 6 I address time as such and, in particular, Bergson's claim that time must be understood exclusively in qualitative terms. I argue that temporal ordering, pastness and futurity cannot be reduced to qualities, and that time cannot be understood without relations. Also, I dispute Bergson's attempt to consider time in separation from space, as there is no purely temporal reality totally free from spatial features.

Chapter 7 marks a transition from duration as a general metaphysical term to its concrete manifestations. Concrete examples of duration, given by Bergson, include psychological and biological processes, movement of a physical body and, as an all-embracing duration that includes all worldly processes, the universe. I propose the duration of a concrete human being as such an all-embracing duration, because a human self involves all layers of being, from minerals to mind, which can acknowledge any worldly phenomena and account for them in an epistemic process. Of course, if the universe could be said to contain all worldly processes, the self merely represents them.

In Chapter 8 I look at epistemic processes and begin to analyse the perception of one's own selfhood in self-consciousness. According to Bergson, an epistemological act is defined either by its analytical or intuitive component, but I contest his opposition of intuition and intellect and present the epistemological act as a three-fold process of primary (pre-conceptual) intuition, intellectual rationalization and secondary (post-conceptual) intuition. I emphasize that the perception of one's self, acquired in this way, gives a picture of an all-embracing unity of human existence, from various manifestations of matter and life to the complexities of mind.

Bergson presents duration as an uncontroversial and harmonious continuity, but Chapters 7 and 8 demonstrate that, inevitably, duration entails discontinuity in various senses. In Chapter 9 I suggest a view on reality which reinstates its continuity. I suggest that when we observe continuity from the past to the present, in actual fact we remain in the present, retaining knowledge of the consecutive events. This knowledge interferes with our view of the past and prevents us from seeing it as a fresh present with an indefinite future. On the other hand, if we look backwards into the past, we can get a sense of continuity, moving from the latest and more complex to the earlier and less complex, without making different temporal periods overlap and interfere with one another.

CHAPTER I

Bergson's Method and Key Concepts

Before we delve into Bergson's texts, a few remarks are necessary to introduce and orientate the reader. In particular, the Bergsonian reader must be
warned against expecting clear definitions – Bergson deliberately avoids
them – and also against accusing Bergson of being inconsistent, as the
change of meaning of his terminology is not an oversight but a result of
this reluctance to commit.

The Creative Mind, Bergson's last book, offers a retrospective view on
the maxims that he set for himself and which he complied with in the earlier
texts. It begins with a criticism of non-Bergsonian philosophy which gives
us an insight into what characteristics, in his opinion, a true philosophy
should be able to demonstrate.

> What philosophy has lacked most of all is precision. Philosophical systems are not cut
> to the measure of the reality in which we live; they are too wide for reality. Examine
> any one of them, chosen as you see fit, and you will see that it could apply equally
> well to a world in which neither plants nor animals have existence, only men, and
> in which men would quite possibly do without eating and drinking, where they
> would neither sleep nor dream nor let their minds wander; where, born decrepit,
> they would end up as babies-in-arms, where energy would return up the slope of its
> dispersion, and where everything might just as easily go backwards and be upside
> down. The fact is that a self-contained (vrai) system is an assemblage of conceptions
> so abstract, and consequently so vast, that it might contain, aside from the real, all
> that is possible and even impossible. (*CM*, p. 11)

Bergson is against systems which are so 'vast' that they can include 'all
that is possible and even impossible' (*CM*, p. 11), and himself strives to get
away from the generalities in favour of concrete and particular things. The
subsequent statement confirms that:

> The only explanation we should accept as satisfactory is one which fits tightly to
> its object with no space between them, no crevice in which any other explanation
> might equally well be lodged; one which fits the object only and to which alone the
> object lends itself. (*CM*, p. 11)

For Bergson, a philosopher must get close to the reality that he endeav-
ours to study, prior to any assumptions as to what that reality might be:
'Metaphysics becomes, in Bergson's hands, a remedial technique in perception, not a form of ethereal contemplation'.[1]

Rather than naming the Bergsonian method 'intuition', as Deleuze
does,[2] for example, I compare it with phenomenological reduction. But
there is one feature which is distinctively Bergsonian: his language stops
where data stops, presenting his readers with a deliberately inconclusive
philosophy, a philosophy where the 'i's are not dotted and 't's are not crossed,
a philosophy where we find ostensive demonstrations rather than final
conclusions and rigid definitions. Striving to gain precision at the level of
description, Bergson is reluctant to commit himself to definitions, because
a definition would dictate what the nature of the phenomenon should be.
He prefers extensive descriptions. As a result, the reader who may have
hoped to find a clear theory of time and self, encounters a seemingly vague
discourse, with changes of meanings throughout the texts, and a refusal to
commit to clear-cut concepts.

In *Mind-Energy* he says of consciousness:

> There is no need to define so familiar a thing, something which is continually pre-
> sent in every one's experience. I will not give a definition, for that would be less clear
> than the thing itself; I will characterize consciousness by its most obvious feature.
> (*ME*, p. 7)

There are other places where he openly refuses to define:

> Neither intelligence nor instinct lends itself to rigid definition: they are tendencies,
> and not things. (*CE*, p. 143)

> Let no one ask me for a simple and geometrical definition of intuition. (*CM*, p. 34)

1   Mullarkey, *Bergson and Philosophy*, 157.
2   See Gilles Deleuze, *Bergsonism* (New York: Zone Books, 1991), Chapter 1, 13–36.

Do not expect of this metaphysics simple conclusions or radical solutions. (*CM*, p. 45)

In the realm of experience ... with incomplete solutions and provisional conclusions, it [metaphysics] will achieve an increasing probability which can ultimately become the equivalent of certitude. (*CM*, p. 46)

This reluctance to commit himself to fixed definitions allowed Bergson to talk of duration in a variety of contexts and made it possible to change the direction of his discourse as necessary. For the purpose of this book, however, a similar reluctance to commit is unnecessary as one can tease out definitions from his discourse. It could be argued that an attempt to systematize Bergson's philosophy would go against his own intentions. However, despite first impressions, Bergson is not an unsystematic thinker. His philosophy is not a collection of random insights. Rather, his ideas developed naturally over four decades. It is possible to extract and organize his ideas on various topics without doing damage to his philosophy as whole and in fact helping his readers to capture and appreciate its distinctiveness and originality. My objective then is not to systematize an unsystematic discourse, but to organize a systematic but disjointed discussion.

In Diagram 1 I attempt to show the conceptual interlacing between elements of Bergson's philosophy at a glance and further on I list key concepts which appear in this work and which change their meaning as the theory of duration evolves.

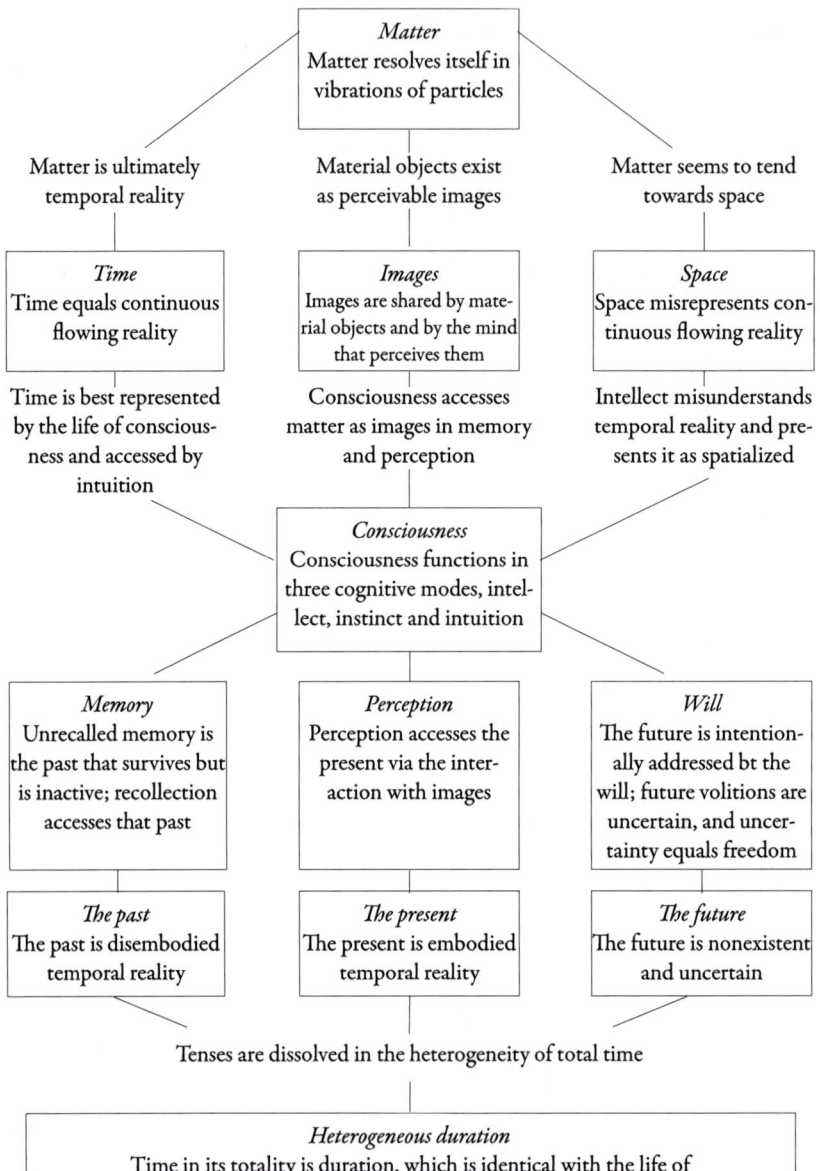

Matter
Matter resolves itself in
vibrations of particles

Matter is ultimately   Material objects exist   Matter seems to tend
temporal reality   as perceivable images   towards space

Time   *Images*   *Space*
Time equals continuous   Images are shared by mate-   Space misrepresents con-
flowing reality   rial objects and by the mind   tinuous flowing reality
                                that perceives them

Time is best represented   Consciousness accesses   Intellect misunderstands
by the life of conscious-   matter as images in memory   temporal reality and pre-
ness and accessed by   and perception   sents it as spatialized
intuition

Consciousness
Consciousness functions in
three cognitive modes, intellect, instinct and intuition

Memory   *Perception*   *Will*
Unrecalled memory is   Perception accesses the   The future is intention-
the past that survives but   present via the inter-   ally addressed bt the
is inactive; recollection   action with images   will; future volitions are
accesses that past   uncertain, and uncer-
                                                                 tainty equals freedom

The past   *The present*   *The future*
The past is disembodied   The present is embodied   The future is nonexistent
temporal reality   temporal reality   and uncertain

Tenses are dissolved in the heterogeneity of total time

Heterogeneous duration
Time in its totality is duration, which is identical with the life of
consciousness, biological life and, ultimately, with all temporal reality
understood as a complex heterogeneity of interdependent elements

Diagram 1 Conceptual framework of Bergson's philosophy

Duration is introduced in *Time and Free Will* as conscious processes, time and motion. In *Matter and Memory*, it is presented as any manifestation of being, either material or spiritual. In *Creative Evolution* it refers to consciousness, concrete manifestations of life, biological evolution and the universe. I accept duration as being in general and as concrete manifestations of being, equating being with temporal reality. I also take on board the Bergsonian idea of *novelty*, whereby each temporally new manifestation of duration is ultimately original and unprecedented.

Heterogeneity is the term that reflects the nature of duration: its elements are individualized and yet bound, so that their identity depends on relations with other elements. Duration perpetually changes and, I claim, not just by adding new elements in the present, but also by modifying its future projection and its past content, and I introduce the term '*constructive retrospectivity*'.

For Bergson, duration is total affirmation, and the *nought* is secondary to affirmation, because before we negate something, we affirm it first. I include the nought in duration as *constructive negation*, whereby sameness of some qualities is opposed to the otherness of other qualities via the quality *x* not being the quality *y*. I also introduce *vectorial properties* into qualities, which are their directedness towards other qualities and their ability to emanate relations with them.

For Bergson, real *time* equals pure duration with all of its elements permeating each other. According to him, regarding it as a homogeneous medium in which elements succeed one another is a false picture of time and is its spatialization. *Space* in *Time and Free Will* appears as an empty homogeneous medium, and in *Matter and Memory*, as an artificial framework. In *Creative Evolution*, the form of knowledge which spatializes is identified as *intelligence*, aimed at the construction of manmade tools. It follows the *cinematographical approach* to reality, taking snapshots of it rather than following its movement. *Instinct*, the form of knowledge aimed at the creation and functioning of organs, grasps duration on the level of biological life, and *intuition*, derived from instinct, can grasp the duration of one's own being and of other objects.

I equate time with temporal reality, but refuse to divorce it from space. Spatial markers, I argue, define and locate duration, and duration itself is

embodied time, events, not pure time and pure motion. As for the epistemological faculties, I assert a three-fold process, beginning with *primary
(pre-conceptual) intuition* – conscious but pre-verbal grasp of sensory data
– continuing with intellectual processing of it, and possibly finishing with
secondary (post-conceptual) intuition which synthesizes what intellect has
provided.

Perception, happening in the present, is movement that provides stimulus for our actions with the participation of the *brain*, which is for Bergson
a conductor and converter of movement, and *affection* in this context means
receiving information about one's body.

The concept of *memory* in Bergson is associated with the surviving
past, with the process of accessing that past and with the recalled past itself.
More specifically, *pure memory* is the past per se when it is not recalled.
When the past is recalled, it becomes either *representational memory* that
translates the past into images or *motor memory* that corresponds to skills,
or, most often, a mixture of both.[3] The role of motor memory in relation
to pure memory is, according to Bergson, such that the former disciplines
the latter, ensuring that only useful recollections are brought into the light
of consciousness. 'Of these two memories that I have distinguished, the
second, which is active, or motor, will, then, constantly inhibit the first, or
at least only accept from it that which can throw light upon and complete
in a useful way the present situation' (*MM*, p. 85).

The process of recalling a memory amounts, for Bergson, to the following sequence: 'Ideas – *pure recollections* summoned from the depths of
memory – develop into *memory-images* more and more capable of inserting themselves into a motor diagram' (*MM*, p. 125).[4] I understand that an
idea (a generalized image) is a vague recollection, and a memory-image (an

3   In *Mind-Energy* Bergson shows the united work of motor memory and representa-
     tional memory. In learning to dance, for example, we use an image ('scheme') which
     is both visual and motor at the same time, and which shows us a new combination
     of the basic motor images which our body is familiar with already (*ME*, 216–17).
4   As noted by Worms, pure recollections are not images, but they are always *given* as
     memory-images (which are not the same as images of perception: memory-images
     are subjective) (Frédéric Worms, *Introduction à Matière et Mémoire de Bergson* (Paris:
     Presses Universitaires de France, 1997), 105.)

image that corresponds to a singular object), is a more precise and definite one. As for a *motor diagram*, it must be a disposition to acknowledge and assimilate the recalled memory-image with a greater focus, 'to the degree that these recollections take the form of a more complete, more concrete and more conscious representation, they tend to confound themselves with the perception which attracts them or of which they adopt the outline' (*MM*, pp. 125–6).

At the end of Chapter 2 Bergson emphasizes his claim that memory and perception are not self-sufficient and complete processes in themselves. He talks of a 'dynamic *progress* by which the one passes into the other' (*MM*, p. 127). An account of this progress is summed up in the following passage.

> On the one hand, complete perception is only defined and distinguished by its calescence with a memory-image, which we send forth to meet it. Only thus is attention secured, and without attention there is but a passive juxtaposing of sensations, accompanied by a mechanical reaction. But [...] the memory-image itself, if it remained pure memory, would be ineffectual. Virtual, this memory can only become actual by means of the perception which attracts it. Powerless, it borrows life and strength from the present situation in which it is materialized. Does not this amount to saying that distinct perception is brought about by two opposite currents, of which the one, centripetal, comes from the external object, and the other, centrifugal, has for its point of departure that which we term 'pure memory'? The first current, alone, would only give a passive perception with the mechanical reactions which accompany it. The second, left to itself, tends to give a recollection that is actualized – more and more actual as the current becomes more marked. Together, these two currents make up, at their point of confluence, the perception that is distinct and recognized. (*MM*, pp. 127–8)

Attention, despite being an effort that brings together pure memory and motor memory, reveals the difference between them: even if motor memory stops being purely pragmatic and becomes more cognitive, it is confined to indicating invariable physical, motor schemes. On the other hand, if motor memory brings into pure memory pragmatic elements, pure memory will still retain its depth.[5]

5   Worms, *Introduction à Matière et Mémoire de Bergson*, 123.

Free will, as discussed in *Time and Free Will*, for Bergson amounts to ultimate self-expression, as against a predictable response to circumstances. It is a continuation of the *creative impetus*, which in *Creative Evolution* means the force that creates life, and is the same as *God, life-drive*, or *life*. (The opposite tendency to remain immobile, realized in plants, is *torpor*.)

Matter is defined in *Matter and Memory* as an aggregate of images, and as the source of inertness in *Creative Evolution*. *Image* means not just an object, but a perceivable object. The nature of duration in Bergson is defined by its *rhythm*, the rate and speed of its processes, with matter being the slowest, and consciousness, the fastest duration.

The *present* in *Matter and Memory* is that which is being made or that which is acting, and the *past* is that which has ceased to act. The *future*, as indicated in *Creative Evolution*, is non-existent and the future reality is undetermined until the very moment of its emergence as present. I claim that the past is changeable because consecutive events add to it new dimensions; that the future is predetermined as a direct projection of the current reality but is nonetheless changeable because the current reality changes; and that the present is elusive because of the *temporal shift* within the self, due to our consciousness perceiving the just past reality whilst existing in the new present. Concrete manifestations of duration are summarized in Table 1.

Table 1 Hypostases of duration

Duration in inanimate matter	Internal vibrations maintaining the integrity of the particular type of matter Locomotion of material bodies
Duration in organic matter	Evolutionary movement Life of an organism Instinctive behaviour of an individual organism
Duration in consciousness	Psychological processes Action

Duration in *Time and Free Will*

1 Duration as Succession

In chapter 1 of *Time and Free Will*, Bergson prepares the reader for his views on time and the self by pointing out a problem in our conventional understanding of states of consciousness. 'It is usually admitted that states of consciousness, sensations, feelings, passions, efforts, are capable of growth and diminution' (*TFW*, p. 1). Bergson argues that the terms 'more' and 'less', normally used to describe the intensity of our emotions, imply spatial relations in the way that what is 'more' is greater in magnitude than what is 'less', and can be regarded as a container for the latter. Bergson argues that these terms are not applicable to states of consciousness, since one such state could not serve as a container for another. Psychic states are not extended as they do not occupy a certain portion of space, and we can only talk of them as intensity and cannot possibly regard them in terms of quantity and magnitude. However, when using terms 'more' and 'less', we treat unextended psychic states as if they were extended, confusing intensity, which does not involve any space-occupancy, and extensity, which is identical to a physical body occupying a concrete portion of space. As a result of this confusion, the specific nature of the intensive is ignored – we regard it in terms of the extensive. 'Common sense agrees with the philosophers in setting up a pure intensity as a magnitude, just as if it were something extended' (*TFW*, p. 3), which for Bergson (as we will see later) is the major problem underlying all other epistemological problems in philosophy.[1]

---

1   It is important to note that Bergson does not wish to confirm common sense. Mentioning common sense in *Matter and Memory* (*MM*, 10–11) and in *An Introduction*

What we take to be a change in magnitude in a feeling or a sensation, is, according to Bergson, a change in quality: not, say, a feeling that has increased in intensity, but different feelings altogether. An apparent change in magnitude is nothing but an alteration of the nature of the emotion or a sensation, an addition of new elements, a qualitative progress, as in the following example.

> It could be easily shown that the different degrees of sorrow … correspond to qualitative changes. Sorrow begins by being nothing more than a facing towards the past, an impoverishment of our sensations and ideas, as if each of them were now contained entirely in the little which it gives out, as if the future were in some way stopped up. And it ends with an impression of crushing failure, the effect of which is that we aspire to nothingness, while every new misfortune, by making us understand better the uselessness of the struggle, causes us a bitter pleasure. (*TFW*, p. 11)

Similarly, in the increasing intensity of pity Bergson finds the following stages that constitute a qualitative progress: it is 'a transition from repugnance to fear, from fear to sympathy, and from sympathy itself to humility' (*TFW*, p. 19).

The reason why we ascribe quantity, or magnitude, to conscious phenomena is that the latter are often accompanied by a muscular effort, 'as if intensity were being developed into extensity' (*TFW*, p. 20). Thus the greater the number of muscles that accompany a conscious state, the greater intensity is ascribed to the corresponding conscious state. 'Consciousness, accustomed to think in terms of space and to translate its thoughts into words, will denote the feeling by a single word and will localize the effort at the exact point where it yields a useful result: it will then become aware of an effort which is always of the same nature and increases at the spot assigned to it, and a feeling which, retaining the same name, grows without changing its nature' (*TFW*, p. 26).

---

to *Metaphysics* ('Introduction', 49), he discusses an alternative faculty, good sense in 'Good Sense and Classical Studies' (See Henri Bergson, 'Mélanges', in Henri Bergson, *Key Writings* (New York, London: Continuum, 2002, transl. Melissa McMahon), 353–4). In this lecture he talks of good sense, an inclination to open oneself for the other, and common sense thus appears as a spatialized, restrictive view of reality that impedes the genuine 'spirit of justice' (ibid., 348).

In other words, Bergson believes that we confuse a feeling with the extended area of our body which is involved in our experiencing the feeling. For instance, we may say of fear that it is more or less great by our heart beating more or less fast.

Thus Bergson arrives at a dual definition of intensity. On the one hand, it reflects the idea of extensive magnitudes coming from without, while on the other hand it reflects the nature of an inner multiplicity, coming from within consciousness itself. Bergson claims that whereas an externally, spatially posited multiplicity contains countable units, the inner multiplicity presents itself as a qualitative diversity and cannot be considered in numerical terms.[2]

The reason for that is that enumeration implies countable units, which must be items, identical in nature but distinct from one another in their spatial position; otherwise they would merge into a single unit. Thus, Bergson argues, the idea of number involves juxtaposition of units, setting them alongside one another – the procedure only being possible in space and not in time. This is introduced by Bergson as the domain where we should consider unextended and intense states of consciousness. This time is presented in *Time and Free Will* as 'pure duration'.

Bergson states, '[T]here are two kinds of multiplicity: that of material objects, to which the conception of number is immediately applicable; and the multiplicity of states of consciousness, which cannot be regarded as numerical without the help of some symbolical representation, in which a necessary element is *space*' (*TFW*, p. 87).

Another important feature of counting juxtaposed objects is the assumption of their impenetrability, which Bergson finds inappropriate in relation to psychic states. 'Feelings, sensations, ideas' in fact 'permeate one another', he observes (*TFW*, p. 89). Our counting distorts their permeating one another, and this also leads to a false understanding of time: 'Now, let us notice that when we speak of *time*, we generally think of a

---

2   Lacey offers a thorough discussion of this, pointing out the necessity to compare
     sensations, which are more or less significant depending on their intensity, and
     difficulties that Bergson's position presents. See A. R. Lacey, *Bergson* (London and
     New York: Routledge, 1993), 3–16.

homogeneous medium in which our conscious states are ranged alongside one another as in space, so as to form a discrete multiplicity' (*TFW*, p. 90).

This serves Bergson as evidence in demonstrating that our view of time is corrupted by space. 'For if time, as the reflective consciousness represents it, is a medium in which our conscious states form a discrete series so as to admit of being counted, and if on the other hand our conception of number ends in spreading out in space everything which can be directly counted, it is to be presumed that time, understood in the sense of a medium in which we make distinctions and count, is nothing but space' (*TFW*, p. 91). 'It follows that pure duration must be something different', Bergson deduces (*TFW*, p. 91). This notion, evolving as his theory develops, at this initial stage, can be understood as a succession of moments as opposed to points in space existing simultaneously.

2   Duration as Movement

As Bergson's theory develops, the notion of duration becomes enriched with extra meaning, and the next idea that is added to duration is the idea of movement. When we perceive oscillations of a pendulum, Bergson says, we do not think of each oscillation as isolated from the preceding one – we perceive them all, 'each permeating the other and organizing themselves like the notes of a tune, so as to form what we shall call a continuous or qualitative multiplicity with no resemblance to number' (*TFW*, p. 105). Just like psychological life, movement too, appears to be pure duration with no spatial features such as divisibility (*TFW*, p. 110).

To reinforce the above assertion, Bergson offers his own solution to the three Eleatic paradoxes, created by Zeno in the fifth century BC. Zeno aimed to prove the totality of being, which, omnipresent and hence immobile, does not allow for movement, as the latter would imply nothingness as a void to be filled.

Zeno's paradoxes

1. Dichotomy

A runner has to run to a given length. Before running the whole length, he must have run half of it. But, before completing the second half, he must have run half of that half, and so on. The division never terminates, as the whole stretch is composed of infinitely many successive pieces. Therefore the runner cannot complete the task, because he cannot finish traversing infinitely many substretches in succession.

2. Achilles

Achilles has to catch up with the tortoise, but he will fail to do so, because as he makes his first step, the tortoise also makes a step, and so on, so Achilles has to traverse infinitely many successive stretches.

3. Arrow

A flying arrow is, in fact, at rest. At any indivisible instant of its flight, the arrow occupies as much space as when it is at rest. Therefore, it is always at rest.

Bergson's response to the above paradoxes is best presented in his own words.

> It is to this confusion between motion and the space traversed that the paradoxes of the Eleatics are due; for the interval which separates two points is infinitely divisible, and if motion consisted of parts like those of the interval itself, the interval would never be crossed. But the truth is that each of Achilles' steps is a simple indivisible act, and that, after a given number of these acts, Achilles will have passed the tortoise. The mistake of the Eleatics arises from their identification of this series of acts, each of which is of a definite kind and indivisible, with the homogeneous space which underlies them. As this space can be divided and put together again according to any law whatever, they think they are justified in reconstructing Achilles' whole movement, not with Achilles' kind of step, but with the tortoise's kind: in place of Achilles pursuing the tortoise they really put two tortoises, regulated by each other, two tortoises which agree to make the same kind of steps or simultaneous acts, so as never to catch one another. Why does Achilles outstrip the tortoise? Because each of Achilles' steps and each of the tortoise's steps are indivisible acts in so far as they are movements, and are different magnitudes in so far as they are space: so that addition

will soon give a greater length for the space traversed by Achilles than is obtained by adding together the space traversed by the tortoise and the handicap with which it started. This is what Zeno leaves out of account when he reconstructs the movements of Achilles according to the same law as the movement of the tortoise, forgetting that space alone can be divided and put together again in any way we like, and thus confusing space with motion. (*TFW*, pp. 112–14)

Bergson's solution demonstrates that duration does not only refer to perception of movement, as this would only mean an example of a psychological state. Such an assumption, however, would be possible and even inevitable, following from Bergson's pendulum example. In connection with this, Bergson stresses the point that 'duration properly so-called cannot be measured' (*TFW*, p. 107), and when we think that we measure duration whilst observing oscillations of the pendulum, we, in fact, measure simultaneities. 'Outside of me, in space, there is never more than a single position of the hand and the pendulum, for nothing is left of the past positions' (*TFW*, p. 108).

At this stage it may seem indeed that the notion of duration is solely applied to the self's perception of movement, and that outside the self there is no duration, which means no movement or time. This is the impression that the reader would get reading the following extract in particular:

> Within myself a process of organization or interpretation of conscious states is going on, which constitutes true duration. It is because I *endure* in this way that I picture to myself what I call the past oscillations of the pendulum at the same time as I perceive the present oscillations. Now, let us withdraw for a moment the ego which thinks these so-called successive oscillations: there will never be more that a single oscillation, and indeed only a single position of the pendulum, and hence no duration. Withdraw, on the other hand, the pendulum and its oscillations; there will no longer be anything but the heterogeneous duration of the ego, without moments external to one another, without relation to number. Thus, within our ego, there is succession without mutual externality; outside the ego, in pure space, mutual externality without succession: mutual externality, since the present oscillation is radically distinct from the previous oscillation, which no longer exists; but no succession, since succession exists solely for a conscious spectator who keeps the past in mind and sets the two oscillations or their symbols side by side in an auxiliary space. (*TFW*, pp. 108–9)

This is superseded by Bergson's treatment of the paradoxes, when he claims that the Eleatics' mistake was based on the misinterpretation of movement. Instead of treating an occurrence of movement as indivisible duration, they confused it with the space traversed and divided that space. That is why they came to a conclusion that, logically, Achilles could not catch up with the tortoise because if he takes a step, the tortoise also takes a step, and so on.[3]

This apparent contradiction between Bergson's own ideas is significant in the way that it, in fact, reveals the development of the idea of duration, which continues throughout all three of the major Bergson texts. As the notion of duration acquires more and more new meanings, comprising ideas of psychic states, time, and movement, we begin to realize that the idea of consciousness and of the self will be connected to these terms as well.

Having laid the foundations for merging time, movement and consciousness, Bergson makes the following claim: 'Pure duration is the form which the succession of our conscious states assumes when our ego lets itself *live*, when it refrains from separating its present state from its former states' (*TFW*, p. 100). In this definition Bergson comes very close to identifying pure duration with the self, as pure duration, pure time uncontaminated with spatial features, is presented as a mode of being for the self in its non-reflective state of simply 'letting itself live'. This involves (1) avoiding being entirely absorbed in the subject matter, 'the passing sensation or idea' (*TFW*, p. 100) and (2) not forgetting the self's former states which should not be set alongside its actual state but merged with it. This way both the past and the present states would form an organic whole similar to the recollection of a tune where notes merge together, producing a heterogeneous unity of melody, while retaining their own individuality at the same time.

---

3   My view is that Bergson considers all types of movement as essentially the same. I am inclined to side with Mullarkey who, aiming to unite the ideas of consciousness, motion and time talks of 'subjectivisation of movement or time' (John Mullarkey, ed., *The New Bergson* (Manchester: Manchester University Press, 2000), 7). For an alternative approach to Bergson, where the internal psychological movement and the external movement in space are distinguished and analysed separately, see Angèle Marietti, *Les Formes du Mouvement chez Bergson* (Paris: Les Cahiers du Nouvel Humanisme, 1953).

According to Bergson, an adequate perception of pure duration is something like this: 'We can thus conceive of succession without distinction, and think of it as a mutual penetration, an interconnection and organization of elements, each one of which represents the whole, and cannot be distinguished or isolated from it except by abstract thought' (*TFW*, p. 101). We would be able to achieve this if we were 'a being who was ever the same and ever changing, and who had no idea of space' (*TFW*, p. 101).

But we tend to spatialize everything, and the above perception of pure duration which Bergson claims to be adequate is contrasted with what he believes to be a false representation of it: 'We set our states of consciousness side by side in such a way as to perceive them simultaneously, no longer in one another, but alongside one another; in a word, we project time into space, we express duration in terms of extensity, and succession thus takes the form of a continuous line or a chain, the parts of which touch without penetrating one another' (*TFW*, p. 101).

The idea of order in succession is also regarded by Bergson as a false representation of a temporal sequence, because by arranging elements in some order we necessarily imagine them as existing simultaneously and thus spatialize the temporal. With the idea of order removed from the temporal domain, the nature of pure duration receives further clarification: 'In a word, pure duration might well be nothing but a succession of qualitative changes, which melt into and permeate one another, without precise outlines, without any tendency to externalize themselves in relation to one another, without any affiliation with number: it would be pure heterogeneity' (*TFW*, p. 104).

3   Space: What Duration is Not

In Bergson's philosophy, heterogeneous duration is contrasted with space which, on the contrary, is 'an empty homogeneous medium' (*TFW*, p. 95). A more detailed definition is this: 'Space is what enables us to distinguish a number of identical and simultaneous sensations from one another; it is thus a principle of differentiation other than that of qualitative differentiation, and consequently it is a reality with no quality' (*TFW*, p. 95).

Bergson is persistent in his effort to expose and counteract our habitual mistake of treating time in spatial terms. 'It is generally agreed to regard time as an unbounded medium, different from space but homogeneous like the latter: the homogeneous is thus supposed to take two forms, according as its contents co-exist or follow one another. It is true that, when we make time a homogeneous medium in which conscious states unfold themselves, we take it to be given all at once, which amounts to saying that we abstract it from duration. This simple consideration ought to warn us that we are thus unwittingly falling back upon space, and really giving up time' (*TFW*, p. 98).

So, whereas space, the medium uniting all beings, exists as a homogeneous medium without quality, time exists as a duration of beings taken as processes. Bergson observes that a common view on time and space is that they are both homogeneous media: in the first one, objects succeed one another and in the second, objects exist simultaneously. Also, we commonly believe that objects that exist in time and space are not outside time and space, but contain time and space within themselves. Bergson distinguishes between time and space as supposed media and time and space occupied and appropriated by objects: the latter are modified space and time, extensity and duration, part of the things' makeup. They appear as much a quality as, say, colour or weight. But whereas space can appear as both medium and extensity of individual objects, time can only be duration of concrete instances of motion and consciousness, and should not be considered as a medium, because this would bring about its inevitable spatialization.[4]

---

4   Naturally, Bergson's treatment of space attracted a lot of criticism. Russell refuses to take Bergson's theory of time and space seriously: 'His doctrine of space is required for his condemnation of the intellect, and if he fails in his condemnation of intellect, the intellect will succeed in its condemnation of him, for between the two it is war to the knife' (Russell, *The Philosophy of Bergson*, 12–13). H. Wildon Carr, by and large sympathetic to Bergson, nevertheless disapprovingly comments: '[I]n this theory the idea of the reality of time excludes the possibility of the reality of space'

4 Duration as Consciousness

Duration appears to be the key term defining conscious phenomena and consciousness per se, but it also refers to time and motion, and we need to find out how time, motion and consciousness participate in each other's existence. Another idea that is added to that of duration in this respect is that of incompleteness: '[I]t is of the very essence of duration and motion, as they appear to our consciousness, to be something that is unceasingly being done' (*TFW*, p. 119).[5]

The primary experiencing of the self in its fundamental and most true form occurs in dreams, or in inattentive perception of sounds. In other words, at a pre-reflective stage, when the self is relaxed and mostly detached from the external world, and is able to just be itself without co-ordinating its fleeting elements with points in the external world. But as soon as we 'wake up', we align the events of our inner life with the external phenomena, breaking duration into segments representing it in the same way as the external elements. In this way we arrive at our habitual understanding of the self and its states as materialized and set out in space. Bergson mentions that there is a gradual passage between the public, spatialized self, and pure

---

(H. Wildon Carr, 'Bergson's Theory of Knowledge', *Proceedings of the Aristotelian Society*, Vol. 9 [1908–1909], 48). Merleau-Ponty points out that to condemn spatialization is unnecessary and insufficient for the comprehension of time (Maurice Merleau-Ponty, *Phenomenology of Perception*, transl. Colin Smith (London: Methuen, 1965), 415).

For a very critical and thorough discussion of time and space in Bergson see S. Alexander, *Space, Time and Deity: The Gifford Lectures at Glasgow 1916–1918*, Vol. 1 (London: Macmillan, 1966).

It will become obvious in later chapters that the relationship between time and space in Bergson is more complex than simply a mutual exclusion. Bergson's refusal to accept time as a medium gives rise to a problem of temporal coexistence (see Chapter 12).

5 H. Wildon Carr criticizes Bergson for attributing various meanings to the concept of duration, but I insist that this is due to the fact that this notion develops gaining new layers of meaning (Carr, 'Bergson's Theory of Knowledge', 45).

duration: the deeper into the self we go, the more vague and fluid states of consciousness are. There are, however, ideas which are not incorporated in the fluidity of our self; ideas that are not generated by the self but brought in from outside our self, offered to us by other people, and which remain within us without being assimilated by us. These are, for instance, commonly held beliefs imposed on us by society as against those born within our own soul (*TFW*, p. 135).

Bergson summarizes his ideas of the self, developed up to this point, in the following terms. We can in principle perceive our conscious life either refracting it through space or directly. If we perceive it directly then we access our own deep-seated conscious states which ought to be understood in terms of quality and not quantity. This means that, although they are qualitatively different, we cannot tell whether this is one state or several, and we cannot individuate them without changing something in their nature. These deep-seated conscious states constitute a duration whose elements form a non-numerical multiplicity and cannot be distinguished from one another. However, this view of one's self is only possible in a hypothetical individual who, immersed in his or her own being, would live a socially isolated life, where society and language would not force him or her to interpret his or her conscious flow as a series of individualized phenomena (*TFW*, pp. 137–8). However, we inevitably spatialize duration under normal circumstances, as 'the intuition of a homogeneous space is ... a step towards social life' (*TFW*, p. 138). Bringing our conscious states outwards, we separate them from one another and express them in words – a precondition for communication.

Thus, as Čapek observed, whereas philosophers were traditionally concerned with subjectivity affecting the account of objective data, Bergson raises the converse issue, namely that of our awareness of introspective data being affected by elements borrowed from sensory experience.[6] Below we shall summarize the characteristics of the true self and of its symbolic representation as they appear in *Time and Free Will*.

6   Čapek, *Bergson and Modern Physics*, 84.

1   The self per se is true duration with psychic states melting in one
    another ('melting' here is the Bergsonian term, which means that
    elements permeate one another but do not become slurred), whereas
    the represented self appears as a collection of separate psychic states.
2   The self is a qualitative multiplicity, pure quality, although it appears
    as a discrete multiplicity and is represented as quantity.
3   The elements of the self are processes, but they are treated as solidified
    objects, external to each other.
4   The elements of the self are never identical, and any new element
    changes the nature of the whole, by virtue of every new element
    happening at a different time. However, our common sense (which
    Bergson disagrees with) regards the elements of the self as identical
    units: it is supposed that one can experience one and the same sensa-
    tion more than once.
5   In its fundamental form the self is locked within itself and is incom-
    municable, because any attempt to externalize psychic events involves
    their juxtaposition and spatialization. This also means that the spa-
    tialization of the self and solidification of its elements aids social life
    because the clarity that the process of externalization entails helps
    communication.
6   As for its mode of existence, duration in its pure form can only exist in
    isolation from other selves, and its spatialized representation is essential
    for social existence, which is communication with other selves.

5   Free Will as Growth of Duration

The primary characteristic of the self as it is depicted in the first two chapters of *Time and Free Will* is that the self in its essence is *dynamic*. The
Bergsonian self does not endure through time preserving some stable core
of its identity – it *is* time in the way that it captures inner processes as they
are and incorporates them into itself. Therefore it is understandable why
Bergson is concerned with the problem of free will in the final chapter, for

the relation of the self to its immediate actions is essential for the dynamic side of the self, for its growth. Stating in the previous two chapters *that* the self is dynamic, in the final chapter Bergson tries to ascertain *how* the self is dynamic. Are its actions predetermined by circumstances or are they free?[7]

In analysing free will, Bergson is not concerned with ethical issues: in *Time and Free Will*, will is solely tied with the problem of the ontology of selfhood.[8] Having demonstrated that the mode of existence for the self is growth consisting in the addition of new psychic states, Bergson is investigating how the growth occurs. Not being interested in the will's externalization as events in the world, as willed action, Bergson is concerned with the psychological state that lies behind the willed action – the making of a decision. The psychic state of willing, we observe, is but one of a multitude of psychic states that the self consists of, but willing, if it is indeed, free willing, would represent the element of growing by itself, as against states such as hunger or sadness caused and controlled by external influences. The question Bergson is faced with is whether our acts of will also belong to the category of externally conditioned states, or whether they are really free and constitute the self's growth from within.

It needs to be emphasized that Bergson's understanding of the freedom of the will is different from the conventional approach: opposing freedom to non-freedom, he associates freedom not with breaking causal links with

---

7   Herman believes that, according to Bergson, 'to become oneself in duration is to be free so that duration and freedom are one and the same thing' (Daniel J. Herman, *The Philosophy of Henri Bergson* (Washington: University Press of America, 1980), 6). I find that although free sprouting is the ideal state for duration, it is not the only way in which it exists and grows: duration can continue in other conscious states, for example, in expectation and contemplation.

8   I agree with Herman when he says that Bergson's freedom 'does not go beyond the self; it neither reaches for the world not for values' (Herman, *The Philosophy of Henri Bergson*, 7). But this is true only in respect of *Time and Free Will*. In *The Two Sources of Morality and Religion*, Bergson applies his ontological findings to the realm of ethics, where ultimately free acts are portrayed as deeds of extraordinary people such as saints. Also, for the ethical implications of Bergson's ideas see Una Bernard Sait, *The Ethical Implications of Bergson's Philosophy* (Whitefish, Montana: Kessinger Publishing, 2005).

preceding phenomena but with unpredictability resulting from radical novelty of genuinely free decisions. Thus whereas non-freedom means for him predictability which is not unlike a rehearsal of the old (rather than determinism), freedom is equated with indeterminacy and a promise of original and new reality.

As for causal links that bring an act of freedom to life, Bergson connects acts of freedom with the emanation of the entire self, in an act of ultimate self-expression, opposing free volitions to predictable automatic actions which are prescribed by the expectations of social demand and are produced mechanically.[9]

The conventional view, as it is understood and criticized by Bergson, and Bergson's own view, are illustrated by Diagram 2 copied from *Time and Free Will*, p. 176.

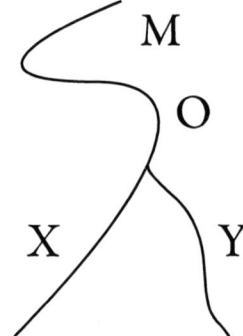

Diagram 2   Emergence of volition

---

9   Bergson's interpretation of the free act as unpredictable and original makes Bachelard say that it is accidental and lacks intellectual causality (Gaston Bachelard, *The Dialectic of Duration*, transl. Mary McAllester Jones (Manchester: Clinamen Press, 2000), 28). In regard to the first remark I must point out that on the contrary, the free act appears to be the conclusion of the existence of a concrete self at the time, uncontaminated by outer circumstances, and in regard to the second remark, that the causal links involved supersede intellectual causality: the driving forces behind the free act include all our unconscious past.

Diagram 2 presents a situation whereby we start an action at the point M and face two possible outcomes, X and Y. At the stage MO we hesitate between directions X and Y, and this very hesitation between two tendencies contributes to the development of the self at the stage MO, 'until the free action drops from it like an over-ripe fruit' (*TFW*, p. 176). However, our common sense which mistakenly understands reality in mechanical terms, as Bergson says, strives to define voluntary activity using clear-cut distinctions. Our mistaken view of this situation is that our consciousness has traversed the series MO, reached the point O and now needs to choose between directions OX and OY which are equally available. In reality, Bergson says, there is hesitation between two tendencies, and neither path is made at the point O. When the choice is made, only one path, say, OX, will thus be created, but our mind will still see the other path as well as an option that we did not choose. Bergson qualifies this as solidification of what is in reality continuous development (*TFW*, pp. 176–7).

The difference between the genuine freedom and its spatialized, mechanical image can be summarized in the following terms.

1 Regarding the path that the self takes whilst making a decision, in genuine voluntary activity the self creates its own path – there is no path, or paths existing beforehand. According to the spatialized view, however, the self follows one of the available paths.

2 Regarding the causal link that connects existing conditions and the act of will, Bergson's view accommodates an idea of continuity of the past and present phases of the self: the self is not independent from the action it is about to take. The decision to take this action and even the predisposition to make this decision contributes to the nature of the self. But the traditional view of the free will, according to Bergson, is that the self, distanced from the situation, decides which option to take.

3 Considering the issue of choice involved in free decisions, Bergson says that there is no neutral activity prior to the moment of making a particular decision – all of the previous life of the self would have contributed towards that decision. He disagrees with the spatialized view, according to which before the self has made the decision, its activity is neutral, so that the decision can be one or the other, before it is made.

Asserting the idea of unpredictability in free decisions, Bergson argues against the very possibility of asking whether it would be possible to predict some future decision, provided we know everything about the person who will make that decision. To have the perfect knowledge of the mental condition of another person would mean knowing, at the deepest level, all of his or her psychic experiences, and also following them at a given moment as any new experiences alter our self by adding new elements to it. Ultimately, having complete knowledge of somebody would mean being that somebody as it would necessarily involve experiencing all his or her experiences. Thus Bergson concludes that it would be absurd to even ask such a question.

6   Automatic Acts

Bergson's account of free acts demonstrates a way in which duration grows, or exists, for 'to exist' and 'to grow' are synonymous for duration. Voluntary acts are primarily growth from within, where the new act is a continuation of all the experiences and memories of all previous acts.

However, our decisions do not always spring solely from our personality. This happens very rarely, and only in critical situations, Bergson says. Most of our actions are a response to some trivial circumstances and do not come from the depth of our being. In our usual occupations, we perform automatic actions resembling reflex acts, engaging the surface of our ego rather than its core.[10]

---

10   Deleuze points at the possibility of the transformation of automatic acts into voluntary acts in Bergson (see Gilles Deleuze, 'Bergson's Conception of Difference', in John Mullarkey, ed., *The New Bergson* (Manchester: Manchester University Press, 2000), 60), but my understanding of this dichotomy is that the two types of psychic activity are mutually exclusive, and that a voluntary act can emerge only through the refutation of automatic acts.

But in this case too, we have to admit that duration exists and grows, being conscious all the time and accumulating new experiences. There is a difference between the two modes of existence and growth, however. Whereas the free act originates solely from the self and in accordance with its inner nature as its very continuation, the automatic act is a response to trivial circumstances and is performed by the superficial strata of the ego and does not touch its depth. In a free act, the growth of duration can be understood as a leap outwards, an *active* externalization of the self in the act. In automatic actions, the growth of the self, as we can see it, is a *passive* accumulation of sensory data and information about one's own acts. In terms of relations with the world, in free actions, the self confronts the established order whereas in automatic actions, the self maintains the established order. As a result, the free self is at risk of being ostracized as its unique and unpredictable actions may threaten social rules, whereas the self that conforms to the rules secures the support of others. In reality, however, duration would oscillate between these extreme patterns of existence and growth.

CHAPTER 3

Body, Soul and the World in *Matter and Memory*

1   Duration as Being

Next Bergson presents duration as being in general, with the distinctive nature of concrete phenomena defined by the specific rhythm of their temporal existence, as a process, and where rhythm could be understood as a rate at which events unroll. The relation between time, consciousness and movement becomes clearer: everything is duration and exists in motion; consciousness is a case of being, understood as duration; what makes consciousness different from other types of being is its specific, very intense rhythm.

In *Time and Free Will* Bergson argued that we misunderstand and immobilize psychological events and movement, which are flowing and indivisible in reality. In *Matter and Memory*, he asserts that this misunderstanding concerns physical reality as well, because everything real is, in fact, in the state of becoming and, consequently, in the state of motion, whereas we always regard the reality as a compilation of solid things (*MM*, p. 191).

Bergson refuses to regard motion as relative, as merely a change of place. Real movement is absolute, he maintains, because 'it emanates from a force' (*MM*, p. 195). One could even interpret Bergson's idea of real movement as an action close to the creation of something new, or as a process close to the act of willing, which could indicate a certain link with consciousness. 'I am assured of the reality of the movement when I produce it, after having willed to produce it, and my muscular sense brings me the consciousness of it. That is to say, I grasp the reality of movement when it appears to me, within me, as a change of *state* or a *quality*' (*MM*, pp. 195–6). Bergson compares the perception of movement with the perception of changes in the qualities of things such as the change from sound to silence, from light to darkness, from one colour to another colour.

He asserts that the material world exists as 'a *moving continuity*', 'in which everything changes and yet remains' (*MM*, p. 197). According to him, this is evident from the fact that in our field of vision we detect no intervals void of coloured objects. 'And, since solids are necessarily in contact with each other, our touch must follow the surface or the edges of objects without ever encountering a true interruption' (*MM*, p. 197). Thus as something necessarily changes in this totality of the real, we must admit that the whole is constantly changing.

However, we follow 'the irresistible tendency to set up a material universe that is discontinuous, composed of bodies which have clearly defined outlines and change their place, that is, their relation with each other' (*MM*, p. 197). The reason why we do that is that 'besides consciousness and science, there is life', and consciousness, which manifests itself in acts, needs to distinguish a 'material zone' that corresponds to its living body (*MM*, p. 198). By analogy with one's own body, others are distinguished as well and as the living being requires nutrition, it is forced to distinguish objects with definite outlines that can serve as food. This subdivision of the real 'answers much less to immediate intuition than to the fundamental needs of life', Bergson observes (*MM*, p. 198), as though in this kind of division we 'prolong the vital movement, but ... turn our back upon true knowledge' (*MM*, p. 199).

Bergson claims that the apparent solidity of matter is due solely to the work of our mind, because 'the preservation of life no doubt requires that we should distinguish, in our daily experience, between passive *things* and *actions* effected by these things in space' (*MM*, p. 200). But, the closer science moves to the ultimate elements of matter, Bergson observes, the farther it moves away from the idea of solidity and discontinuity. Thus he is inclined to regard objective matter as ultimately consisting of the vortices of atoms, energy fields – in a word, as, primarily, motion.

We are getting a picture of the world where everything is defined by motion – in fact, everything is motion – and where differences between various fragments of reality amount to the differences of rates at which processes, comprising the given durations, unroll. These rates, faster or slower ones, are defined by the speed at which vibrations, comprising the movements of duration, take place. This intensity of movement of each being defines its relations with other beings:

> May we not conceive, for instance, that the irreducibility of two perceived colors is due mainly to the narrow duration into which are contracted the billions of vibrations which they execute in one of our moments? If we could stretch out this duration, that is to say, live at a slower rhythm, should we not, as the rhythm slowed down, see these colors pale and lengthen into successive impressions, still colored, no doubt, but nearer and nearer to coincidence with pure vibrations? In cases where the rhythm of the movement is slow enough to tally with the habits of our consciousness – as in the case of the deep notes of the musical scale, for instance – do we not feel that the quality perceived analyses itself into repeated and successive vibrations, bound together by an inner continuity? (*MM*, p. 203)

A particular quality, then, is determined by the rhythm of vibrations that constitute the flow of duration of the given object (here Bergson means vibrations of the ultimately small particles of the body). Thus we may conclude that if duration is movement, it is not a uniform process but one which comprises repetitive mini-processes whose amplitude, constant each time, defines the outer view of the object perceived for the perceiving mind which lives its life in line with its own rhythm of duration. We can understand that, according to Bergson, all objects should be in themselves identical to each other at the level of a single vibration; it is the number of vibrations per unit of time that makes things look different, of whose objectivity Bergson says: 'Motionless on the surface, in its very depth it [matter] lives and vibrates' (*MM*, p. 204).[1]

---

1   Commentators are baffled by Bergson's apparent attempts to eliminate substance for motion. See, for example, Russell, *The Philosophy of Bergson*, 18–19; Lacey, *Bergson*, 101; Marietti, *Les Formes du Mouvement chez Bergson*, 31; Carr, 'Bergson's Theory of Knowledge', 45. In *The Creative Mind*, however, Bergson explains that he does not deny substance but analyses it in terms of movement, referring to 'things in the making' rather than 'things made' (*CM*, 188). He also remarks: 'Let me insist I am ... in no way setting aside *substance*. On the contrary, I affirm the persistence of existences' (*CM* [hardback], 305).
   I am inclined to interpret Bergson's earlier apparent anti-substantialism as a claim that substance exists but is never at rest and always changing. Density and motion

As for human consciousness, Bergson describes it here as 'a duration with its own determined rhythm, a duration very different from the time of the physicist, which can store up, in a given interval, as great a number of phenomena as we please' (*MM*, p. 205). Bergson exemplifies it by comparing the perception of red light, which lasts one second, and 400 billion physical vibrations of waves that constitute red light and which correspond to one second of psychological perception: it would take at least 25,000 years if these vibrations were to be perceived separately, separated by 0.002 seconds, necessary to distinguish them (*MM*, pp. 205–6).

Bergson maintains that the independent existence of a red light and our perception of it constitute different durations with different rhythms, with consciousness contracting the enormous number of physical vibrations into one moment of conscious life. We can point out, though, that whatever inner differences there are in the intensity of their rhythms, both our perception of the red light and the red light emitting 400 billion successive vibrations, happen simultaneously and both last the same length of time equal to one second. So the temporal difference must lie in the speed at which the phenomena unroll in our consciousness and outside it, and not in the actual length of the processes involved.

Bergson continues: 'In reality there is no one rhythm of duration; it is possible to imagine many different rhythms which, slower or faster, measure the degree of tension or relaxation of different kinds of consciousness and thereby fix their respective places in the scale of being' (*MM*, p. 207). Within this context, Bergson says about perception that 'to perceive consists in condensing enormous periods of an infinitely diluted existence into a few more differentiated moments of an intenser life, and in thus summing up a very long history. To perceive means to immobilize' (*MM*, p. 208).

----

thus constitute one and the same reality of matter, and the universal movement of movements is not a void, but a process where the element of volume and density and the element of motion are both necessary predicates of material reality, and neither one nor the other can be abstracted from it without destroying the integrity of matter. Substance cannot be imagined still, a-temporal and motionless, nor can motion be imagined empty of volume. Both components are inseparable from each other. For more on this point see Chapter 7, Section 2.

Thus the radical difference between mind and matter seems to amount to the difference in intensity and concentration of the movement involved in their makeup: the mind, able to contract in what is our present the multitude of events and conceive them in their connection with each other, is able to join them in a meaningful unity. 'My perception ... contracts into a single moment of my duration that which, taken in itself, spreads over an incalculable number of moments' (*MM*, p. 208).

But, unperceived, matter exists as it is, relaxed, and 'sensible qualities, without vanishing, are spread and diluted in an incomparably more divided duration' (*MM*, p. 208). And here we arrive at Bergson's ultimate definition of matter per se: 'Matter resolves itself into numberless vibrations' (*MM*, p. 208), where the particles that vibrate are so small that the solidity of the object apparent at a macro-level, disappears when we start talking of atoms and electrons.

2   Bond of Mind and Matter in Perception

By declaring everything to be duration, Bergson lays grounds for the unity of matter and mind as versions, or parts, of one and the same being, and by introducing diverse rhythms, he tries to account for the difference between them. His next task is to investigate how they come together in human existence. As we have seen from Bergson's example of perceiving a colour, when fragments of being with different rhythms affect each other, they translate the rhythm of another being into their own. According to Bergson, in perception we impose our own, very intense, rhythm of duration onto the reality which exists and moves at a different, much more diluted and relaxed rate so that what appears to our eye to be a simple and brief event is, in itself, a multitude of events. If we could free ourselves from the organizing power of our consciousness, we would be able to distinguish all of the micro-events that constitute what seems to us to be a single one.

Material objects do not merely exist in themselves, undetectable and imperceptible, but can be heard, felt, seen and smelt. Bergson finds this a decisive feature of physical bodies and defines matter as 'an aggregate of images' (*MM*, p. 9). As Bergson says, 'by "image" we mean a certain existence which is more than that which the idealist calls a *representation*,[2] but less than that which the realist calls a thing – an existence placed halfway between the "thing" and the "representation" … [T]he object exists in itself, and, on the other hand, the object is, in itself, pictorial, as we perceive it: image it is, but a self-existing image' (*MM*, pp. 9–10).

The concept of image is complicated. However, it seems that by introducing this term Bergson tries to dissolve the subject-object ontology. Placing an image between a representation and a thing, Bergson supplies a term, which explains the continuity of the object and the subject. Whereas representation belongs exclusively to the subject as a private mental process, and a thing is a fragment of physical reality with no reference to the subject, an image is a feature that belongs to both: the perceivable object is an image for the subject, and the subject accesses and appropriates this image.

An image belongs to the mind because it is what we find in ourselves when we see, hear or touch: pictures, sounds and sensations. But it also belongs to the object inasmuch as the latter appears at the superficial level as a picture, sound and touch. If our perception is to be compared with a photograph of things, then we must realize that this photograph 'is already taken, already developed in the very heart of things' (*MM*, p. 38).[3]

---

2   The translators of *Matter and Memory* Nancy Margaret Paul and W. Scott Palmer point out that représentation ('representation') which is, in French, very often used to mean 'perception', is used throughout the book as meaning a mental picture (*MM*, 251).

3   A more detailed exposition and interpretation of images in Bergson, close to our position, can be found in F. C. T. Moore, *Bergson: Thinking Backwards* (Cambridge: Cambridge University Press, 1996), 23–32. For a clearer understanding of the issue, it may be useful to compare and contrast Bergson's theory with alternative accounts of images and imagination. Whereas for Bergson, imagery is the result of the filtering of reality, Crowther's analysis, for example, emphasizes the creative aspect of imagination and the ability of images to be detached from the immediacy of their origin. (Paul Crowther, 'Imagination and Objective Knowledge', in Paul Crowther, *Philosophy*

In the light of Bergson's claim that everything exists as movement, perception too must be understood as a process. In particular, Bergson refuses to treat perception as a faculty of acquiring knowledge for the sake of knowledge (*MM*, p. 28). Instead, he invites us to regard perception as a phase of the interaction between the subject and the object, whereby perception informs the subject of the range of one's possible actions connected with the object. 'We note ... that a strict law connects the amount of conscious perception with the intensity of action at the disposal of the living being' (*MM*, p. 31). Presumably, Bergson has in mind that organisms' fields of perception vary, and this coincides with the limitations of their acting power, whereby bodies receive as much information as they can utilize. The diversity of knowledge available to the individual is correlative with the range of his or her choice of action. Sight and hearing allow us to perceive objects found at a distance which, according to Bergson, can serve as a measure of indetermination that surrounds the activity of a living being as it indicates the range of objects that the living being can contact deliberately.

Given the utilitarian character of perception, we must also accept that the object is perceived according to our potential interest in it. In Bergson's terms, this amounts to the fitting of the perceived object into the stream of continuous movement which would involve both the object and the subject. However, the adaptation of the object for perception does not involve its modification or the addition of any new elements but is a selection of the existing ones. 'There is for images merely a difference in degree, and not of kind, between *being* and *being consciously perceived*' (*MM*, p. 37).

---

after Postmodernism: Civilized Values and the Scope of Knowledge (London and New York: Routledge, 2003), 66–77, especially 73–5). Husserl opposes imagination and perception (Edmund Husserl, *The Idea of Phenomenology*, transl. William P. Alston and George Nakhnikian (The Hague: Martinus Nijhof, 1964), 54), and J. J. C. Smart refuses to believe that images exist at all (J. J. C. Smart, 'Mind and Brain', in Richard Warner and Tadeusz Szubka, eds, *The Mind-Body Problem: A Guide to the Current Debate* (Oxford: Blackwell, 1997), 20). For an explanation of various usages of the term 'imagination' see Gilbert Ryle, *The Concept of Mind* (London: Hutchinson, 1969), 245–79.

The object as it is perceived is less than the object itself because 'being bound up with all other images, it [the object] is continued in those which follow it, just as it prolonged those which precede it' (*MM*, pp. 35–6). In other words, what we perceive is the superficial image taken out of the context of the relations it has with other objects. So, 'a *present* image' is the real object taken in the totality of its existence and 'a represented image' is an extraction of what is relevant to our potential action.[4]

Bergson says, for a present image to become a represented image, 'it would be necessary, not to throw more light on the object, but, on the contrary, to obscure some of its aspects, to diminish it by the greater part of itself, so that the remainder, instead of being encased in its surroundings as a *thing*, should detach itself from them as a *picture*' (*MM*, p. 36). To reinforce this view, Bergson makes the following explicit declaration.

---

4   The Bergsonian version of perception generated the following criticism. Russell accuses Bergson of confusing the mental act of knowing with the material thing which is known and blurring the distinction between mind and matter (Russell, *The Philosophy of Bergson*, 21). By saying that, Russell is plainly ignoring the key Bergsonian idea that, albeit different in nature, both mind and matter are processes and that perception is the area of their fusion. Alexander, on the other hand, accepts this: 'In pure perception ... contact between subject and object is absolute to the extent of transcending spatial and temporal relations wholly' (I. W. Alexander, *Bergson: Philosopher of Reflection* (London: Bowes and Bowes, 1957), 38). Moore questions what exactly we do select in perception and Lacey struggles to individuate images in Bergson: if two people can perceive the same wind as both hot or cold, it is puzzling when Bergson says that images exist without being perceived (Lacey, *Bergson*, 90). Lacey's query, leading directly to the issue of primary and secondary qualities, can be easily resolved: a concrete image, as I understand, is the making of the subject and object, where the subject picks up that aspect of the object which is relevant to its needs and senses, and so the wind entails more than just one image, and can be perceived as both hot and cold by different subjects: an image is 'an object in its specific relation to a particular body capable of action' (Moore, *Bergson: Thinking Backwards*, 31). Regarding primary and secondary qualities, I agree with Mullarkey's explanation: 'The qualities of sounds, colours, tastes and smells are objective ... in principle if not in fact, not because they pre-exist their actual sensuous state in some virtual form (for that would assume that they are not the product of omission), but because perception begins with the object de *jure*' (Mullarkey, *Bergson and Philosophy*, 44).

'We are too much inclined to regard the living body as a world within a world, the nervous system as a separate being, of which the function is, first, to elaborate perceptions, and, then, to create movements. The truth is that my nervous system, interposed between the objects which affect my body and those which I can influence, is a mere conductor, transmitting, sending back or inhibiting movement' (*MM*, pp. 44–5).

Bergson shows that we learn about our self through perceptions (the images that we know from without) which become affections (the images that are delivered from within one's body) (*MM*, p. 17). Our body which we do not identify with our self and our personality, is, nevertheless, a means by which we gain evidence about our own being. Through our body, so to speak, we learn about our self from outside, and this serves us as the first step towards our apprehension of our inner nature.

Perception necessarily becomes affection since no influence from outside can reach our mind except via our body. Having analysed disturbances we receive from light and sound, Bergson remarks: '[T]he sensations here spoken of are not images perceived by us outside our body, but rather affections localized within the body' (*MM*, p. 52). Bergson affirms: '[T]here is no perception without affection', characterizing affection as 'the impurity with which perception is alloyed' (*MM*, p. 58).

In connection with this he offers his version of the transitory connection from matter to mind stating: '[W]e pass by insensible degrees from the representative state, which occupies space, to the affective state which appears to be unextended' (*MM*, p. 52). Affective states, we learn, are 'vaguely localized' and are 'intermediate states' between images and ideas, 'the former extended and the latter unextended' (*MM*, p. 53).[5]

Bergson detaches perception and affection as processes different in degree, placing perception outside one's body, and affections within it. 'Just as external objects are perceived by me where they are, in themselves and not in me, so my affective states are experienced where they occur, that is, at a given point in my body' (*MM*, p. 57).

---

5   Whereas image is a singularity that corresponds to each individual perception, idea is a generalization and corresponds to a type of a certain image, not to a concrete perception.

Thus the continuity of existence is seen by Bergson in the following terms. Physical reality enters our mind as images by means of perception and affection, and the existence of the real is continued in our mind where physical things are stripped of their physicality but are retained as images.[6] The role of our mind consists not in confronting reality as a subject but prolonging it in a de-materialized form as known and, later, as remembered.

3   Duration as Memory

As I have observed, the principal role of perception in Bergson is to provide a stimulus for one's actions. But the action effected in response is not the only possible action but the action chosen by the individual as the most suitable under the given circumstances. The possibility of making a choice would require more than just sensory data presented to the mind; the mind must be able to know of possible options, which would be based on the previous knowledge that the individual has accumulated about the world, as 'this choice is likely to be inspired by past experience, and the reaction does not take place without an appeal to the memories which analogous situations may have left behind them' (*MM*, p. 65). Therefore perception does not exist in its pure form as described above – real perception necessarily involves memory: '[W]hen perception, as we understand it, is once admitted, memory *must* arise' (*MM*, p. 43). The role of memory in perception is clearly defined: 'We assert, at the outset, that if there be memory, that is, the survival of past images, those images must constantly mingle with our perception of the present and may even take its place'

6   Here the image that belongs to both subject and object in perception becomes diluted temporally: its material base and its existence as a mental picture belong to different times.

(*MM*, pp. 65–6).[7] Our perception, Bergson argues, is always enriched by memories to such an extent that 'perception ends up by being merely an occasion for remembering' (*MM*, p. 66).[8]

Although perception and memory 'always interpenetrate each other' (*MM*, p. 67), they are qualitatively different processes. Perception corresponds to the actuality of the present whereas memory belongs to the domain of the past. In regard to action and movement, perception is active and memory is not: 'the past is only idea, the present is ideo-motor' (*MM*, p. 68), and Bergson's definition of the past and the present is this: 'The past is essentially *that which acts no longer*', and the present is '*that which is acting*' (*MM*, pp. 68–9).

Chapter 2 of *Matter and Memory* is an attempt to reveal how new content, delivered to our consciousness via perception, becomes welded onto the previous content via the faculty of memory. Bergson's theory of memory comprises three theses.

1) The first thesis concerns two forms of memory: 'The past survives under two distinct forms: first, in motor mechanisms; secondly, in independent recollections' (*MM*, p. 78).

Bergson uses the example of learning a lesson to demonstrate the presence of two distinct forms of memory in our daily actions. He distinguishes between (1) remembering the content of a lesson, an achievement which has occurred as a result of repeating the lesson several times, and (2) remembering each successive instance of repeating the lesson. He termed the first

---

7   To compare Bergson's views with contemporary findings in memory research, see Endel Tulving and Fergus Craik, eds, *The Oxford Handbook of Memory* (New York: Oxford University Press, 2000).

8   For a brief overview of ideas relating to the division of memory into primary memory (awareness of the present) and secondary memory (recollection of the past), with references to William James, Hebb, Shannon and Weaver, Broadbend, Atkinson and Shiffrin, as well as to Braddeley and Hitch working memory model, see Jackie Andrade, ed., *Working Memory in Perspective* (Hove, UK: Psychology Press, 2003), 5–17.

type of memory 'motor memory' and the second, 'pure recollection'. The characteristics of the two types of memory are summarized below.

1 The apparent difference between pure recollection and motor memory is that the memory of learning a lesson is remembered as an event – it is unique and cannot be reproduced. Motor memory, on the other hand, is the ability to recite the lesson by heart.[9]

2 Pure recollection is a representation and as such it is grasped intuitively: in pure memory, we imagine. Motor memory is like a habit: it requires repetition and effort. It is stored in a motor mechanism which is set in motion by an impulse; in motor memory, we repeat.

3 A memory image recalled in pure memory can be remembered instantaneously in its entirety or in part. On the other hand, motor memory requires a definite time for all movements to unroll as one remembers the content of the lesson. It is an action rather than a representation, and as such it is a part of one's present rather than of one's past. It can only be realized when it is lived and acted.

4 Whereas pure recollection represents our past to us, conserving images of the past, motor memory acts our past in our present, prolonging the effect of past images into the present.

5 Recollections occur involuntarily, effortlessly and spontaneously, as against motor memories which are acquired voluntarily and with an effort.

6 Recollections have a fixed date in our personal history; they are unique and cannot be repeated, as each new recollection is a different event. Motor memories are taken out of their temporal co-ordinates and are impersonal; they are essentially built by repetitions.

7 Bergson's view is that only pure memory deserves to be considered as true, genuine memory and that motor memory is a habit interpreted

---

9   For a useful summary of the key features of habit memory in Bergson and an illuminating link with the Merleau-Pontian study of body in *The Phenomenology of Perception* see Edward S. Casey, 'Habitual Body and Memory in Merleau-Ponty', *Man and World* (1984), 17: 278–97, and a very brief but to the point illustration of habit memory can be found in Moore, *Bergson: Thinking Backwards*, 37–8.

by memory rather that memory itself. It is only by mistake that it has been taken as true memory in psychology.

8 The two types of memory do not function in isolation but constantly interact: pure memory records memory images of all events of our life. As every perception is prolonged in a potential action, the movements which accompany pure memory, modify the organism and create in it new dispositions toward action. The disposition toward action facilitates the formation of new mechanisms of action which are activated in the present and directed to the future.

The key role of pure memory in relation to motor memory, according to Bergson, is such that the latter disciplines the former, ensuring that only useful recollections are brought into the light of consciousness. 'Of these two memories that we have distinguished, the second, which is active, or motor, will, then, constantly inhibit the first, or at least only accept from it that which can throw light upon and complete in a useful way the present situation: thus, as we shall see later, could the laws of the association of ideas be explained' (*MM*, p. 85).

2) The second thesis concerns recognition in general, and, in connection with this, memory-images and movements.[10] 'The recognition of a present object is effected by movement when it proceeds from the object, by representations when it issues from the subject' (*MM*, p. 78).

Recognition is described by Bergson as 'the concrete process by which we grasp the past in the present' (*MM*, p. 90). It seems as though, Bergson begins, that first there occurs perception. Then we search our memory for

10 A more recent discussion of images in memory and non-image representation for single visual-special inputs can be found in W. F. Brewer and J. R. Pani, 'The Structure of Human Memory', in G. H. Bower, ed., *The Psychology of Learning and Motivation: Advances in Research and Theory*, Vol. 17 (New York: Academic Press, 1983), 1–38. The authors raise the issue of difficulty in creating the mental 'video recording' of a complex event. For a discussion of repeated events, generic event images and generic non-image representation, see R. C. Schank and R. P. Abelson, *Scripts, Plans, Goals and Understanding* (Hillsdale, New Jersey: Erlbaum, 1977).

a similar image which would aid perception and result in recognition. But 'in most cases recollection emerges only after the perception is recognized', the philosopher observes (*MM*, p. 91), because we recognize, at the same time as we perceive, without necessarily having in our mind similar images of the past.

Bergson distinguishes, first of all, 'an *instantaneous* recognition, of which the body is capable by itself, without the help of any explicit memory-image. It consists in action and not in representation' (*MM*, pp. 92–3). An example of such recognition, as it appears in *Matter and Memory*, is an automatic action of moving about in a familiar place, as opposed to (say) hesitating at every corner in a strange town.

> Now, between these two extremes, the one in which perception has not yet organized the definite movements which accompany it and the other in which these accompanying movements which are organized to a degree which renders perception useless, there is an intermediate state in which the object is perceived, yet provokes movements which are connected, continuous and called up by one another. (*MM*, p. 93)

Thus we shift from the state where our conscious awareness is at its highest to the state where our conscious awareness is superfluous, depending on our familiarity with the surroundings.

Bergson sees a close connection between motor reaction and recognition. 'At the basis of recognition there would thus be a phenomenon of a motor order', he concludes, because 'to recognize a common object is mainly to know how to use it' (*MM*, p. 93).

3) The third thesis is about the gradual passage of recollection into movement, about recognition and attention: 'We pass, by imperceptible stages, from recollections strung out along the course of time to the movements which indicate their nascent or possible action in space. Lesions of the brain may affect these movements, but not these recollections'. (*MM*, p. 79)

Bergson proposes the hypothesis that memories, just like perceptions, are neither the function of the brain nor are they contained inside it. He maintains that movement can produce only movement, and that stimulation of the brain can only produce 'a certain attitude into which recollections will come to insert themselves' (*MM*, p. 99) and that brain damage affects only the person's ability to access recollections and not recollections themselves since the latter are not found in the brain.

> Sometimes they [lesions of the brain] would hinder the body from taking, in regard to the object, the attitude that may call back its memory-image; sometimes they would sever the bond between remembrance and the present reality; that is, by suppressing the last phase of the realization of a memory – the phase of action – they would thereby hinder the memory from becoming actual. But in neither case would a lesion of the brain really destroy memories. (*MM*, p. 99)[11]

Naturally, the reader of Bergson would want to ask: if memories are not located in the brain, *where* are they located? But as it is clear from Chapter 3 of *Matter and Memory*, Bergson dismisses the legitimacy of such a question. Memory is itself the past, and we cannot ask where something which is past is. Physical location is a characteristic of things present, since 'the past is only idea, the present is ideo-motor' (*MM*, p. 68). Spatiality and physicality are the features of actuality, and it is what a thing loses by becoming past.

---

11   Clinical cases of recovering memories after amnesia following a head injury could be used to illustrate Bergson's point that if memories were stored in parts of the brain, they would be lost due to the physical damage to those parts. See, for example, Herbert F. Crovitz, 'Loss and Recovery of Autobiographical Memory after Head Injury', in David C. Rubin, ed., *Autobiographical Memory* (Cambridge: Cambridge University Press, 1988), 273–89, for one complete case of successful memory retrieval. However, contemporary literature also supports a sceptical position that such recovered memories are not strictly speaking remembered but confabulated. For examples of confabulation of memories after an injury see Alan Baddeley and Barbara Wilson, 'Amnesia, autobiographical memory, and confabulation', in David C. Rubin, ed., *Autobiographical Memory* (Cambridge: Cambridge University Press, 1988), 225–52; Nelson Butters and Laird S. Cermak, 'A Case of Study of the Forgetting of Autobiographical Knowledge: Implications for the Study of Retrograde Amnesia', in Rubin, ed., *Autobiographical Memory*, 253–72; For an illustration that pseudo-memories occur when subjects are under pressure to recall the past, see J. Dywan and K. Bowers, 'The Use of Hypnosis to Enhance Recall', *Science*, 222, 184–5.

The process of recalling a memory amounts, for Bergson, to the following sequence: 'Ideas – pure recollections summoned from the depths of memory – develop into memory-images more and more capable of inserting themselves into a motor diagram' (*MM*, p. 125). We understand that idea (a generalized image) is a vague recollection, and a memory-image (an image that corresponds to a singular object), is a more precise and definite one. As for a motor diagram, it must be a disposition to acknowledge and assimilate the recalled memory-image with a greater focus, 'to the degree that these recollections take from the form of a more complete, more concrete and more conscious representation, they tend to confound themselves with the perception which attracts them or of which they adopt the outline' (*MM*, pp. 125–6).

At the end of Chapter 2 Bergson emphasizes his claim that memory and perception are not self-sufficient and complete processes in themselves. He talks of a 'dynamic *progress* by which the one passes into the other' (*MM*, p. 127). An account of this progress is summed up in the following passage.[12]

12   Russell accuses Bergson of confusing 'the present occurrence of a recollection and the past occurrence which is recollected' and, forgetting that perception and recollection are both present facts, claiming to have accounted for the difference between the present and the past, and then, confusing the present cognitive act of remembering with the past object of remembering, blurring the distinction between past and present (Russell, *The Philosophy of Bergson*, 21). I would agree with Russell that Bergson uses the term 'memory' in more than one sense. According to Endel Tulving, 'memory' has the following meanings: (1) neurocognitive capacity to encode, store and retrieve information; (2) a hypothetical store in which the data are accumulated; (3) information in that store; (4) property of that information; (5) retrieval of that information; (6) phenomenal awareness of remembering something. (Endel Tulving, 'Concepts of Memory', in Endel Tulving and Fergus Craik, eds, *The Oxford Handbook of Memory* (New York: Oxford University Press, 2000), 36). Bergson denies (2), ignores (6) and (4), and throughout his texts applies the term 'memory' to either (1), (3), or (5) without warning, which may give an impression that these meanings are interchangeable. However, despite this shortage of suitable terminology and contrary to Russell, Bergson does not lose the thread of his argument, and one can clearly understand the different meanings of the term according to the context.

On the one hand, complete perception is only defined and distinguished by its coalescence with a memory-image, which we send forth to meet it. Only thus is attention secured, and without attention there is but a passive juxtaposing of sensations, accompanied by a mechanical reaction. But, ... the memory-image itself, if it remained pure memory, would be ineffectual. Virtual, this memory can only become actual by means of the perception which attracts it. Powerless, it borrows life and strength from the present situation in which it is materialized. Does not this amount to saying that distinct perception is brought about by two opposite currents, of which the one, centripetal, comes from the external object, and the other, centrifugal, has for its point of departure that which we term 'pure memory'? The first current, alone, would only give a passive perception with the mechanical reactions which accompany it. The second, left to itself, tends to give a recollection that is actualized – more and more actual as the current becomes more marked. Together, these two currents make up, at their point of confluence, the perception that is distinct and recognized. (*MM*, pp. 127–8)

4   The Spatiality of Time and the Temporality of Space

We realize that duration, although depicted in *Time and Free Will* as purely inextensive and spiritual, could not exist as such in reality because its existence presupposes growth, the addition of new content, and the material for new content which comes from the physical world. Data about it is delivered to the consciousness by our senses so that we experience the very materiality of the world – we partake of it as recipients of the direct influence that weight, size, colour or sound produce on our body. In this sense, we shall have to conclude, duration partakes of the materiality and extensity of the physical world since its content comprises our sensations originating from a physical cause.

In Chapter 4 Bergson narrows the problem of the relation between the inextensive and the extensive to the very point where their union occurs – the human body. If in *Time and Free Will* duration was presented as pure intensity, pure temporality, considered separately from anything spatial and physical, in *Matter and Memory* Bergson explicitly talks of one's body

participating in the formation of duration. Body, as Bergson puts it, 'represents the actual state of my becoming, that part of my duration which is in
process of growth' (*MM*, p. 138). Later he adds: 'It is in very truth within
matter that pure perception places us, and it is really into spirit that we
penetrate by means of memory. But, on the other hand, while introspection reveals to us the distinction between matter and spirit, it also bears
witness to their union' (*MM*, p. 180).

Bergson talks about 'reconciliation between the unextended and the
extended' (*MM*, p. 181). This reconciliation takes place in pure perception:
'We place the perceived images of things outside the image of our body,
and thus replace perception within the things themselves. But then, our
perception being a part of things, things participate in the nature of our
perception. Material extensity is not, cannot any longer be, that composite
extensity which is considered in geometry; it indeed resembles rather the
undivided extension of our own representation. That is to say, the analysis of
pure perception allows us to foreshadow in the idea of *extension* the possible
approach to each other of the extended and the unextended' (*MM*, p. 182).

What the above observation seems to mean is this. It is clear from our
every day experience that matter (the extended) and the mind (the unextended) constantly interact in perception. The mind is not some entity
locked within itself and impenetrable by physical influences. On the contrary, we observe that it is, in fact, open to influences from the material
world. And this means that (1) they must have some common basis which
would allow interaction between them, and (2) there must be a channel
through which the work of interaction is effected or a mediating agency
that effects the communication between them.

We observe that our mind is aware of the physical qualities of things
in such a way that it does not just know *of* them – our mind has a direct
experiential knowledge of physical things. Unextended, it experiences their
extensity in sensations delivered to it by the nervous system of the body.
Thus, extensity and inextensity must partake of each other: inextensity as
a recipient of the physical influence and extensity, as a bearer or producer
of a visual, audio or tactile image that can be detached from it mentally.
This is how we interpret Bergson's position regarding the unity of spirit

and matter occurring in perception.[13] Pure memory, too, should help us in this reconciliation: 'Our conception of pure memory should lead us, by a parallel road, to attenuate the second opposition, that of quality and quantity' (*MM*, p. 182).

Bergson believes that he has found an answer to the problem of spirit and matter: there is a gradual passage or transition between them. The 'function of spirit is to bind together the successive moments of the duration of things', and 'it is by this that it comes into contact with matter and by this also that it is first of all distinguished from matter', and if the above assertions are correct, then 'we can conceive an infinite number of degrees between matter and fully developed spirit – a spirit capable of action which is not only undetermined, but also reasonable and reflective. Each of these successive degrees, which measures a growing intensity of life, corresponds to a higher tension of duration and is made manifest externally by a greater development of the sensory-motor system' (*MM*, p. 221).

Bergson takes a further step in trying to reconcile spirit and matter by demonstrating that there is no difference in nature between them. In Summary and Conclusion of *Matter and Memory* the universe itself is declared to be a kind of consciousness: 'The material universe itself, defined as the totality of images, is a kind of consciousness, a consciousness in which everything compensates and neutralizes everything else, a consciousness of which all the potential parts, balancing each other by a reaction which is always equal to the action, reciprocally hinder each other from standing out' (*MM*, p. 235).

What makes an individual consciousness different is that it continues and retains the past in the present and escapes the law of necessity, 'the law which ordains that the past shall ever follow itself in a present which merely repeats it in another form and that all things shall ever be flowing away' (*MM*, p. 235).

---

13   Alexander suggests that, in Bergson, one's body is both the object and the subject, when one knows it from inside, which then presumably makes it the domain where the object and the subject are merged (Alexander, *Bergson: Philosopher of Reflection*, 31).

Bergson summarizes his view on the mind / body problem as 'the threefold opposition of the inextended and the extended, quality and quantity, freedom and necessity' (*MM*, p. 244). The explanation of the union of spirit and matter thus amounts for him to 'suppressing or toning down these three oppositions' (*MM*, p. 244).

1   The inextended and the extended.
'That which is given, that which is real, is something intermediate between divided extension and pure inextension. It is what we have termed the *extensive*' (*MM*, p. 245).
The extensive participates both in things (when they are taken out of an abstract space) and in our consciousness as we perceive extensity. Also, extension admits of degrees, which implies degrees between the inextended and the extended.[14]

2   Quality and quantity.
The opposition of quality and quantity is the same, Bergson asserts, as that of consciousness and movement, inasmuch as the latter represents a series of calculable changes. Concrete movement is in a way like consciousness, since it can prolong its past into its present. Also, movement, like consciousness, can give rise to sensible qualities, Bergson claims, as it is the rhythm of our duration that contracts diluted reality into distinct sensible qualities. Thus the idea of *tension* of a particular movement helps Bergson to overcome the opposition between quality and quantity.

3   Freedom and necessity.
Absolute necessity, which is a property of matter, consists, for Bergson, in 'a perfect equivalence of the successive moments of duration, each to each' (*MM*, p. 247). However, the duration of the material universe does

14   Mark Antliff makes an interesting suggestion that, since Bergson distinguishes matter and spirit as phenomena with different rhythms, it is rhythm that is a connecting link between extensity and intensity (Mark Antliff, *Inventing Bergson: Cultural Politics and the Parisian Avant-Garde*, (Princeton, New Jersey: Princeton University Press, 1957), 100).

not follow this pattern of existence precisely. The material universe allowed the apparition of living beings, 'capable, even in their simplest forms, of spontaneous and unforeseen movements' (*MM*, p. 248). Thus necessity and freedom are present in the universe in the form of mutual dependency. 'Freedom always seems to have its roots deep in necessity and to be intimately organized with it. Spirit borrows from matter the perceptions on which it feeds and restores them to matter in the form of movements which it has stamped with its own freedom' (*MM*, p. 249).

Bergson attempts to reinforce his claim that the integral connection of spirit and matter can be explained by showing that they both originate ontologically from the same source and share the same basic feature: they both partake of motion to such an extent that motion becomes their primary characteristic.

We understand that the passage between spirit and matter is effected in concrete instances of perception where image plays the crucial role of an intermediary element that belongs to both subject and object. What Russell understood in decidedly negative terms as the 'confusion of subject and object' in Bergson,[15] we interpret as the purposeful weakening of the traditional opposition of subject and object. Whereas the traditional opposition is based on the subject and object being separate entities, standing on different ontological platforms, on the subject confronting the object and on that fact that the epistemological contact between them is a subjective process taking place in the subject's mind, Bergson demonstrates by the use of the term 'image', that the epistemological act, involving both the subject and the object, is an objective process. Instead of the opposition of subject and object we are presented with two objects, one of which generates an image, and the other which perceives it, and the ontological roles of the perceiving mind and perceived matter are levelled out. Both members of the former opposition become objectified, and perception is no longer a private process, hidden in one's mind, but an externalized process, the flow between two objective positions which were formerly known in philosophy as subject and object. Synonymizing 'subject' with 'mind' or 'spirit',

15   Russell, *The Philosophy of Bergson*, 23.

we de-subjectify the inextensive, and deprive it of its mysterious, private character. Open to the objective process of perception, mind becomes merely a destination point for imagery emitted by the extended object.

5   Continuity of the Past, Present and Future in Memory, Perception and Action

Comparing the past and the present, and correlating them with memory and perception, Bergson finds that the past and the present have a qualitative difference: 'There is much more between past and present than a mere difference of degree. My present is that which interests me, which lives for me, and in a word, that which summons me to action; in contrast, my past is essentially powerless' (*MM*, p. 137).

The real, concrete present, Bergson continues, is necessarily of some length.

> What is, for me, the present moment? The essence of time is that it goes by; time already gone by is the past, and we call the present the instant in which it goes by. But there can be no question here of a mathematical instant. No doubt there is an ideal present – a pure conception, the indivisible limit which separates past from future. But the real, concrete, live present – that of which I speak when I speak of my present perception – that present necessarily occupies a duration. (*MM*, p. 137)

So, 'what I call "my present" has one foot in my past and another in my future' (*MM*, p. 138).[16] Inasmuch as it is in my immediate past, it is a sensation, and in the sense of it being my immediate future, it is action or movement. 'Whence I conclude that my present consists in a joint system of sensations and movements. My present is, in its essence, sensory-motor'

---

16   For an example of a complete theory of continuous present, see Andros Loizou, 'The Dynamic of Time', in Andros Loizou *Time, Embodiment and the Self* (Aldershot, UK: Ashgate, 2000), 22–56.

(*MM*, p. 138). What this means is, of course, that he is talking about the content that fills our being in a present situation, which consists in 'the consciousness I have of my body' (*MM*, p. 138).

An attempt could be made to keep body and duration apart by saying that what constitutes growth of duration in the present is not sensations and actions themselves, heavily loaded with materiality, but our own inner thoughts and feelings accompanying sensations and actions. But all of the preceding discussion on perception and memory indicates that Bergson takes the opposite direction. He attempts to bring materiality into the life of the mind, to incorporate, so to speak, space into time. We read on:

> In that continuity of becoming which is reality itself, the present moment is consti-
> tuted by the quasi-instantaneous section effected by our perception in the flowing
> mass, and this section is precisely that which we call the material world. Our body
> occupies its centre; it is, in this material world, that part of which we directly feel the
> flux; in its actual state the actuality of our present lies. If matter, so far as extended
> in space, is to be defined (as we believe it must) as a present which is always begin-
> ning again, inversely, our present is the very materiality of our existence, that is to
> say, a system of sensations and movements and nothing else. And this system is
> determined, unique for each moment of duration, just because sensations and move-
> ments occupy space, and because there cannot be in the same place several things at
> the same time. (*MM*, p. 139)

In *Time and Free Will*, the self was described as ever-growing duration which appeared as a heterogeneous but not numerical multiplicity of all the psychic states one has ever experienced. Acts of volition which added their content to duration were the only examples of its growth. In *Matter and Memory*, we are presented with growth of duration via perception enriched by memory. Perception is the conductor of new content being incorporated into our conscious life, and memory is the previously acquired content which gives meaning to what is new. An act of volition, in this context, is the process that follows perception and memory as a stage preceding action.

Thus we can understand that the process of growth is of a cyclical nature, each cycle comprising perception, recognition, processing informa-tion, volition and action. All these stages present a continuity of the inex-tensive past flowing into the material present and looking into the future.

Perceptive knowledge is impossible without the involvement of the physical world. On its own, duration cannot produce material for its growth – it needs material from outside. Matter reaches our mind via the agency of our body. Via the senses, we become acquainted with colours, shapes, weights and sizes, and we feel that this acquaintance is direct to the extent that we ourselves partake of matter – not identifying ourselves with the material object, but as direct recipients of its influence resulting from its combination of physical properties.

Memory aids perception in the form of recognition which occurs as soon as perception has taken place. Our past experiences are in our mind in a dormant state, and as soon as the process of perception begins, they anticipate the further moment of perception and rush to our mind in order to accompany perception as it progresses. Bergson insists that in remembering, we literally return to the past: '... the truth is that we shall never reach the past unless we frankly place ourselves within it' (*MM*, p. 135).

The willed decision is made at this stage – the decision that seems to be a leap in our conscious life, something radically new with an element of creation ex nihilo. However, it is possible that this decision is but a summary of the process of remembering and recognizing where possibilities had already presented themselves to our mind as various combinations of memory-images and actuality. These possibilities could have appeared before our mind as potential decisions and it is only with a slight effort on our part that we choose one of them to become actualized.

As far as our conscious awareness of action is concerned, it occurs as a direct effect of our act of volition by means of our body and with restrictions imposed on it by physical laws. In relation to matter affected by our actions and movements, our decision acts as a kind of a physical law – a law with unique application, perhaps, but obeyed automatically by the material object just as any physical law would be obeyed.

In action we, of course, perceive, and so we come back to stage one of what we called the cycle of growth of duration. This cycle of growth constitutes the existence of duration in the present. It is the present of duration where present perceptions, revived past memories and physical movement directed towards the future are actualized in their full intensity. By contrast, what is purely past in duration, is inextensive, powerless and does not interest our mind. It is pure memory, for Bergson: memory in the state when it is not remembered.[17]

But we remember that duration comprises all of the psychic experiences which we have ever had. We have summarized the mode of performance for duration at the time of its growth, the present moment, which follows from the past and looks into the future but the whole of duration must also include those experiences that we do not explicitly remember. Also, because every component of duration is connected with every other component of that duration, the past which is not remembered must still be connected with our present state which contains present perceptions and recalled memories.

Even when we do not remember most of our past, it nevertheless enters our every action and every thought because the manifestation of our self at every moment is a result of the totality of our past experiences. For when we make a decision and engage ourselves in action, what we decide and what we are conscious of in our action is but a fraction of what is actually happening when we act.

When carrying out an action, we have voluntary control over the pursuit of an end. But much of our act is not under our control – the manner of our speech, the way we walk, feelings we have – all these components just happen as we engage in the realization of our plans and must represent that very past that we do not remember but which nevertheless forms part of our duration, of our self.

Bergson comes to reconsider the role of the unremembered past and, even though in Chapter 2 he treats the unremembered past as inactive, in Chapter 3, he admits that 'our character, always present in all our decisions, is indeed the actual synthesis of all our past states. In this epitomized form our previous psychic life exists for us even more than the external world, of which we never perceive more than a very small part, whereas, on the contrary, we use the whole of our lived experience' (*MM*, p. 146).

---

17 For a more recent discussion of conscious (explicit) and unconscious (implicit) memory, see, for example, Stephan Lewandowsky, John C. Dunn, Kim Kirsner, eds, *Implicit Memory* (Hillsdale, New Jersey: Lawrence Erlbaum Associates, 1989).

Further on he writes: 'The whole of our past psychical life conditions our present state, without being its necessary determinant; whole, also, it reveals itself in our character, although none of its past states manifests itself explicitly in character' (*MM*, p. 148).

Whereas our waking existence gravitates towards perception, matter and the present, in dreams we become liberated from the pressure of the present situation. Sleep, Bergson observes, breaks the connection between the necessity of the actual and our memories. When we dream, 'memories, which we believe abolished, then reappear with striking completeness; we live over again, in all their detail, forgotten scenes of childhood; we speak languages which we no longer even remember to have learned' (*MM*, pp. 154–5).

The opposite to a person in a sleeping state would be the impulsive personality of someone who lives 'only in the present' and responds 'to a stimulus by the immediate reaction which prolongs it' (*MM*, p. 153). The normal self, Bergson points out, never stays in these extreme positions – it adopts intermediate positions, borrowing from memory what is useful for the present action.

Referring to the problem of the selection of images, Bergson does not believe that images, present in our mind, trigger other images and bring them forth to our consciousness. He insists that the totality of our memories, even those not remembered, is present for us, and it is by expanding our entire consciousness which explores its own depths, that the access to memory-images is effected. Our consciousness oscillates between the actuality and the domain of pure memory, resulting in 'an infinite number of possible states of memory' (*MM*, p. 168).

CHAPTER 4

Consciousness, Life and the Universe in *Creative Evolution*

1   Duration as the Universe

In Chapter 1 of *Creative Evolution*, Bergson compares psychological duration with inanimate things and living bodies. Existing as a human being for Bergson is, first of all, passing from state to state, from being warm to being cold, from working to doing nothing. 'I change, then, without ceasing' (*CE*, p. 1). Our states constantly change even if only by passing through time and becoming an instant longer. Consequently, 'there is no essential difference between passing from one state to another and persisting in the same state' (*CE*, p. 2).

Duration is irreversible because it consists of an accumulation of our ever-growing experiences, and this makes it impossible for any state to occur more than once. A person now is not the same as he or she was a moment ago and therefore cannot have the same physical state as in the past. So, 'for a conscious being, to exist is to change, to change is to mature, to mature is to go on creating oneself endlessly' (*CE*, p. 8).

Whereas change for psychological duration is an accumulation of memories, Bergson maintains, for material objects it is 'a displacement of parts which themselves do not change': atoms (*CE*, p. 8). Change here is reversible: 'When a part has left its position, there is nothing to prevent its return to it. A group of elements which has gone through a state can therefore always find its way back to that state, if not by itself, at least by means of an external cause able to restore everything to its place. This amounts to saying that any state of the group may be repeated as often as desired' (*CE*, pp. 8–9). Nothing new is created in the realm of the material world and what 'will be is already present in what it is' (*CE*, p. 9).

In this way Bergson assigns to duration and to inanimate things distinct modes of existence: in duration, change is an accumulation of memories, but for material objects, it is a displacement of unchangeable parts; duration ages, has a history and therefore it is irreversible, whereas in material objects change is reversible and repeatable; also, every moment brings novelty to duration, and its future is unforeseeable, whilst in material objects there is no novelty and the future is calculable.

But what about a living organism? A material object does not grow, but a living body does. The body's current state is to be explained by all of its past and its heredity as against the material object, whose present moment depends exclusively on the previous one. This means for Bergson that a living body, unlike a material object, does have duration. 'Continuity of change, preservation of the past in the present, real duration – the living being seems, then, to share these attributes with consciousness', Bergson observes (*CE*, p. 24).

Bergson becomes more and more convinced that 'organic evolution resembles the evolution of a consciousness, in which the past presses against the present and causes the upspringing of a new form of consciousness, incommensurable with its antecedents' (*CE*, p. 29). Like consciousness, evolution is unpredictable and, like consciousness, life is creating something new at every moment.

Refusing to associate a living, organized body with an object, an unorganized body, Bergson nevertheless finds it appropriate to compare the living organism to 'the totality of the material universe', founding this comparison on 'the essential character of organization' common to both the living organism and the universe itself: they both grow and endure; their 'past, in its entirety, is prolonged into present, and abides there, actual and acting' (*CE*, p. 16).[1]

---

1   In line with Bergson's thought which treats the universe as an organism albeit discriminating between organisms and things on the macro level, Russell observes that the radical difference between living matter and inanimate matter is that the former is chemically dynamic and the latter, chemically static but also that if, distinguishing evolution from mere change, we define it as an 'increase of complexity and heterogeneity', then 'there is reason to believe that there has been evolution also

Besides, despite the initial opposition of duration versus material objects, we are beginning to see that reality in general, and not just the reality of the self and the living organisms in particular, is understood as a process rather than a static thing, with time being an integral part of the makeup of each process. But, as Bergson says, obeying our intellect in dealing with reality, we freeze in our mind the flowing eventuality of the world and present it to ourselves as solid spatialized segments.

It seems at first as though, for Bergson, the difference in the mode of being of duration and material objects affects their temporality in the following way: duration lives its time and the latter has absolute significance for it, whereas for an object, its length does not matter. As Bergson asserts, time is absolute for duration, but for objects as we see them, it is a relation between the elements involved. However, we later realize that what Bergson means is that it is merely *our view* of the inanimate reality and that we, in order to help our understanding of the world, distort reality, real systems, and construct artificial systems which do not reflect the actual state of affairs and are a product of our intellect.

Real processes take some real, concrete time to happen, and these periods of time can only be regarded as absolute because they cannot be contracted or stretched. The latter is, in our view, one of the key statements of Bergson's philosophy of time. He illustrates it by giving an example of sugar being dissolved in a glass of water.

Even in the material world, he says,

> history ... unfolds itself gradually, as if it occupied a duration like our own. If I want to mix a glass of sugar and water, I must, willy-nilly, wait until the sugar melts. This little fact is big with meaning. For here the time I have to wait is not that mathematical time which would apply equally well to the entire history of the material world, even if that history were spread out instantaneously in space. It coincides with my impatience, that is to say, with a certain portion of my own duration, which I cannot protract or contract as I like. It is no longer something thought, it is something lived. What else can this mean than the glass of water, the sugar, and the process of

in the inanimate world', given the development of the solar system which 'accounts admirably for the development of galaxies' (Bertrand Russell, *Human Knowledge: Its Scope and Limits* (London: Routledge, 1992), 45–6).

the sugar melting in the water are abstractions, and that the Whole within which they have been cut out by my senses and understanding progresses, it may be in the manner of a consciousness? (*CE*, p. 10)

Obstacles which lie in the way of the philosophy of duration involve two illusions, as stated in Chapter 4 of *Creative Evolution*: the idea of the original void which stands between consciousness and duration, and the cinematographical nature of our perception which means that we grasp snapshots of reality at intervals.

All action is directed at an absence or a void, Bergson remarks, either getting something we want or creating something that does not yet exist. Whereas he finds this way of looking at things legitimate in the sphere of action, he declares it a mistake to apply this view to the sphere of ontology.

An attempt to annihilate objects leaves us with an idea of, at least, a place that they had occupied, which is in itself something positive. If we think of an object as non-existing rather than annihilated, we still think of an object as existent first, so thinking of it as non-existent is adding the idea of exclusion to the idea of an object. Hence Bergson's assertion that '*there is more, and not less, in the idea of an object conceived as "not existing" than in the idea of this same object conceived as "existing"; for the idea of the object "not existing" is necessarily the idea of the object "existing" with, in addition, the representation of an exclusion of this object by the actual reality taken in block*' (*CE*, p. 302). This amounts to saying that the idea of the non-existent is not negative enough.

Bergson disagrees with the view that being has a beginning and that initially there was a complete void, later replaced by being. If we pass from the idea of the nought to the idea of being, being is presented as a logical or mathematical, non-temporal essence: it is a static conception of the real where everything is given once and for all, in eternity. Being appears as a logical construct because one cannot observe the continuity of nought and being in reality. Bergson invites us to see being directly, without the imagined nought coming between us and the real being, so as to preserve the continuity of being and prevent the need to explain the inexplicable leap from nothing to something (*CE*, p. 315).[2]

2   For more on the nought see Chapter 5, Section 2.

As for the cinematographical view of reality, Bergson's considerations are as follows. Before we distinguish bodies around us, we distinguish qualities, such as colours, sounds, resistance etc (*CE*, p. 317). They present themselves to us as immobilities succeeding each other, but close analysis will confirm that they are nothing but vibrations, so that ultimately, 'every quality is change' (*CE*, p. 317), because what we regard as immobile matter is, in fact, a continuous movement. Unable to represent moving reality as moving, we take snapshots of it and string them onto an abstract idea of becoming. Thus Bergson compares the mechanism of our knowledge to the cinematograph: both attempt to reconstruct movement out of immobilities, perceived from outside, whereas movement is indivisible and is articulated inwardly, from within. Taking the same cinematographical view regarding transitions within the development of a human being, we talk of childhood, adolescence and adulthood as if they are stops, ignoring the transition from one to another, thereby creating a wrong picture of the self (*CE*, p. 329).[3]

2   Duration as Evolution

Satisfied that a conscious process is similar in essence to a living process of a body, Bergson begins a search for their common ontological origin, which, complementing his theory of images, leads eventually to the merging of ontology and epistemology, where *knowing* is regarded as part of *being*.

---

3   Bergson here could have been inspired by the early history of filmmaking. For example, Muybridge arranged to take photos of a moving horse and used the magic lantern to project the sequence of still images presenting the horse as galloping on the screen (Erik Barnouw, *Documentary: A History of the Non-fiction Film* (Oxford: Oxford University Press, 1993), 3). Most importantly, he was able to present the horse as racing at various speeds, the feature which may have given rise to the Bergsonian idea that moving reality in cinema and in perception in general is manipulated to the point of breaking it down to still images, and reconstructing motion at will and not as it really is.

Bergson suggests that we should regard '*life in general*' as a 'current', flowing from one body to another and from one generation to the next, rather than an abstraction, 'a mere heading under which all living beings are inscribed' (*CE*, p. 27). This current is seen by the philosopher as ever intensifying as it advances, organizing bodies and dividing itself amongst species and individuals: '*Life is like a current passing from germ to germ through the medium of a developed organism*' (*CE*, p. 28). 'The essential thing is the *continuous progress* indefinitely pursued, an invisible progress, on which each visible organism rides during the short interval of time given it to live' (*CE*, pp. 28–9).[4]

Turning to the theories of evolution developed by the time of his writing of *Creative Evolution*, Bergson rejects radical finalism, which implies a reality where 'things and beings merely realize a programme previously arranged' (*CE*, p. 41). If there is nothing unforeseen, no invention or creation in the universe, time is useless, for predetermination, in Bergson's view, devalues the time that it actually takes for an evolutionary process to unroll.

The idea of a pre-existing model is contrary to the reality enduring through time, Bergson claims, where there is no ontological foundation for repetition, for 'if everything is in time, everything changes inwardly, and the same concrete reality never recurs' (*CE*, p. 48). Everything that emerges in the living process is new, and what we see as repetitions has been extracted from reality by our mind. '[W]e do not *think* real time. But we *live* it, because life transcends intellect' (*CE*, p. 49).

For our acts we can always find antecedents that caused them, such as intentions, Bergson observes. 'In this sense mechanism is everywhere, and finality everywhere, in the evolution of our conduct' (*CE*, p. 50). But that is an external view of our inner life, because intention aims at nothing but rearranging the past whereas action is a new reality, which the intellect can 'resolve indefinitely into intelligible elements without ever reaching

4   The present state of the study of evolution is represented by the synthetic theory which accepts mutations as raw material for evolution and natural selection which determines the direction of evolution. See, for example, Julian Huxley, *Evolution: A Modern Synthesis* (London: George Allen and Unwin, 1942).

its goal' (*CE*, p. 50). Bergson claims to propose a philosophy of life which deals with evolution itself rather than with its result – the evolved.

Bergson's theory presents the organized world as a harmonious whole. But for Bergson, harmony exists not as a fact but as a principle: different forms of life conflict with each other, because each individual retains a certain impetus from the universal vital impulsion and uses its energy to further its own adaptation. Thus harmony, in the shape of complementarity, appears only in tendencies. Also, this harmony is not an end and is not in front of us – it is rather behind 'due to an identity of impulsion and not to a common aspiration' (*CE*, p. 54). Bergson does not believe that life pursues an end. While the process of evolution is taking place there is only a direction to follow, but no final destination. We can talk of an end only retrospectively, tracing back the development that led to a particular result so that there is an end as a result but not as a pre-existing, i.e. already given, model.

Bergson needs to find the principle responsible for the evolutionary changes which ultimately constitute the evolutionary progress. He remarks that adaptation of an organism to the outer conditions consists in replying to those conditions rather than repeating them and this, the philosopher warns, implies an intelligent or quasi-intelligent activity in evolution. Therefore, he concludes, we must introduce into evolution the idea 'of an *original impetus* of life, passing from one generation of germs to the following generation of germs through the developed organisms which bridge the interval between the generations' (*CE*, p. 92). It is this impetus – élan – that causes the variations which create new species. Élan has a psychological dimension, though for Bergson 'psychological' does not necessarily mean 'anthropomorphic'. It is a cosmological force, which is not tied to human form.

Bergson maintains that in order to understand the work of evolution, we ought to consider the creation of an organ, the complexity of its structure and the simplicity of its function. These two aspects belong to different levels of reality. Whereas 'the simplicity belongs to the object itself, ... the infinite complexity [belongs] to the views we take in turning around it, to the symbols by which our senses or intellect represent it to us, or, more generally, to elements *of a different order*, with which we try to imitate it artificially, but with which it remains incommensurable, being of a different nature' (*CE*, pp. 94–5). For instance, whereas nature created an eye in one simple act, we conceive this act as a complex feat of manufacturing, consisting of a multitude of stages. Bergson maintains that in reality nature's act of constructing the eye is as simple and effortless as our raising our hand: it is one undivided process that has no parts.[5]

3   Human Consciousness as a Fragment of Evolution

Bergson's claim that the evolutionary movement is the movement of duration means to us that firstly, an individual duration and the evolutionary movement are modelled in the same way, and secondly, that an individual duration, which does not appear ex nihilo each time a person is born, must be a prolongation of the evolutionary movement. In particular, Bergson is interested in the principal direction of the evolution that leads to man. 'Our main business is to determine the relation of man to the animal kingdom, and the place of the animal kingdom itself in the organized world as a whole' (*CE*, p. 111).

According to Bergson, two causes explain the diversity of the evolutionary development: 'the resistance life meets from inert matter, and the explosive force – due to an unstable balance of tendencies – which life bears within itself' (*CE*, p. 103). Unlike an individual who has to choose between different tendencies, as he or she cannot realize them all, nature preserves different tendencies and 'creates with them diverging series of species that will evolve separately' (*CE*, p. 105). However, only one tendency has been fully developed in nature – the one that leads through the vertebrates up to a man.[6]

---

5   Mullarkey makes a useful point that Bergson, unlike Richard Dawkin, is against 'graduism' in evolution (Mullarkey, *Bergson and Philosophy*, 71).

6   Bergson's discussion of the human evolution must have been fuelled by Haeckel's exemplification of evolutionary development founded on the claim that the

Bergson suggests distinguishing lines of evolution not 'by the possession of certain characters, but by [their] tendency to emphasize them' (*CE*, p. 112). Vegetables and animals can be distinguished by their method of alimentation: a vegetable derives its nutrients, carbon and nitrogen, directly from the air, water and soil; an animal can assimilate these elements only after being primarily absorbed by plants. Vegetables do not need the ability to move whereas animals, unable to assimilate directly the carbon and nitrogen, must be able to move as they seek vegetables and/or other animals for their nourishment. Therefore 'animal life is characterized, in its general direction, by mobility in space' (*CE*, p. 114). Fixity and mobility are characteristics of vegetables and animals respectively but, Bergson points out, they can only be regarded as prevalent tendencies, as there are cases of moving and insectivorous plants and immobile, parasitic animals.

Thus a connection between duration, conscious activity and movement, introduced in *Time and Free Will*, is explained in *Creative Evolution* by their mutual dependence: organisms able to perform complex movements demonstrate a higher degree of consciousness.[7] 'The humblest organism is conscious in proportion to its power to move *freely*' (*CE*, p. 117).

Movement, namely, free movement of a living organism, appears as a prolongation of the conscious activity of the individual, and this correlates with another line of Bergson's thought where he maintains that individual conscious activity is a prolongation of the work of evolution. Thus everywhere we find movement, which is modified as it passes through different media, but nevertheless preserves its continuity throughout these processes.

The final distinction that Bergson makes between plants and animals is that the former are unconscious and the latter are conscious. 'To sum

---

development of the embryo (ontogeny) is a speeded-up replay of the evolution of the species (phylogeny). See Ernst Heinrich Phillip Haeckel, *The Evolution of Man*, Vol. 1 and 2 (Whitefish, Montana: Kessinger Publishing, 2004).

7   Bergson sees consciousness even in the organisms that do not have a developed nervous system. He believes that the nervous system arises from a division of labour and takes the conscious function to a higher level, but does not create it itself. 'It would be as absurd to refuse consciousness to an animal because it has no brain as to declare it incapable of nourishing itself because it has no stomach' (*CE*, 116).

up, the vegetable manufactures organic substances directly with mineral substances; as a rule, this aptitude enables it to dispense with movement and so with feeling. Animals, which are obliged to go in search of their food, have evolved in the direction of locomotor activity, and consequently of a consciousness more and more distinct, more and more ample' (*CE*, p. 118).

Bergson suggests, 'The first living organisms oscillated between the vegetable and animal form, participating in both at once' (*CE*, p. 118). Then the two lines emerged by following different tendencies: one of mobility and consciousness, another, of fixity and insensibility. The very cause of life at its origin is 'an effort to engraft on to the necessity of physical forces the largest possible amount of *indetermination*' (*CE*, p. 121).

The direction in which higher organisms developed is marked by the increased performance of the structures supporting the nervous system, Bergson continues. The latter is inserting indetermination, unforeseeability, into matter which is a manifestation of the primary role of life at the level of a concrete individual. A nervous system then is in itself a *'reservoir of indetermination'*, the ultimate result of the work of the vital impulse (*CE*, p. 133).

The three directions of the vital impulse, vegetative torpor, instinct and intelligence, are all mutually complementary and mutually antagonistic. Plants have some mobility dormant within them and animals are sometimes prone to torpor. The same works for instinct and intelligence, which 'having originally been interpenetrating, retain something of their common origin' (CE, p. 142). Human intelligence, Bergson maintains, is, above all, manifested by man's manufacturing ability, a materialized invention. But there are some animals which can construct a crude instrument (apes) or recognize a constructed object, such as a trap (foxes), and this demonstrates elements of intelligence in animals. Diagram 3 illustrates Bergson's interpretation of the evolutionary development, where the original impetus of life, containing three different tendencies (vegetative torpor, instinct and intelligence), divided itself into plants and animals.

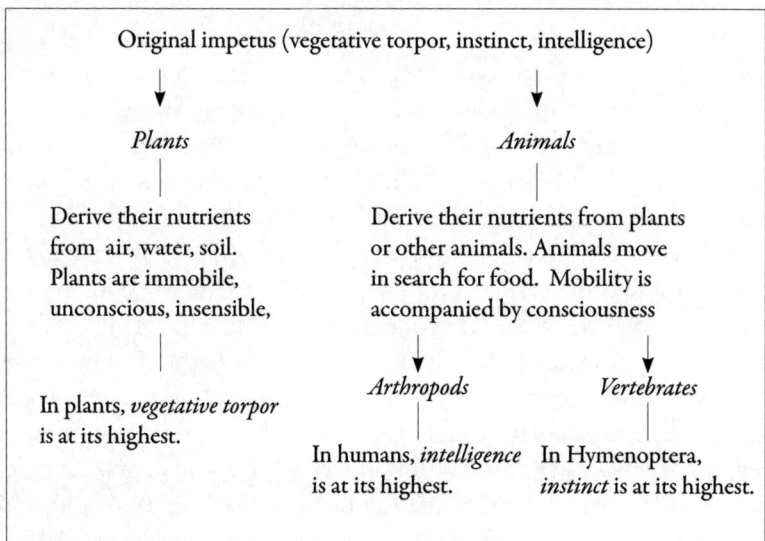

Diagram 3 The evolutionary development of consciousness

As the evolutionary development results from life's explosive force plus resistance of matter, Bergson asserts a firm link between the evolution of life and the work of consciousness: evolution unrolls as if consciousness had penetrated matter, loading it with possibilities which are slowed down by the resistance of matter. 'Consciousness launched into matter' is thus a new angle that Bergson gives to the definition of life (*CE*, p. 191).

As for the relation between mind and things, Bergson regards 'the intellect as a special function of the mind, essentially turned towards inert matter' (*CE*, p. 217). Intellect and matter adapted to each other in order to fulfil the objective of the sentient beings to gain control over matter. '*This adaptation, has, moreover, been brought about quite naturally, because it is the same inversion of the same movement which creates at once the intellectuality of mind and the materiality of things*' (*CE*, p. 217).

During the course of the evolution, the original impetus was divided by matter to accompany the plurality of individual beings, unity and plurality being categories of inert matter. Bergson asserts that at the origin of life there is supra-consciousness, an explosive creative tendency which lies dormant until the creation becomes possible. In this context, individual consciousness appears to be a mere fragment of supra-consciousness that, in its perpetual motion, is passing through concrete individuals.

What differentiates us from animals qualitatively is that an animal brain can only set a limited number of motor mechanisms whereas our brain is unlimited in this respect. Consciousness corresponds to the living being's power of choice, being aware of possible actions that surround the real action, and is synonymous with invention and with freedom. Bergson admits that man is the end of evolution because in every other life form consciousness has found its limit, but man continues the vital movement indefinitely. Having said that, man is not the ideal result of the evolutionary process. Bergson would have preferred to see in humans not just perfected intellect but equally developed intuition: 'A complete and perfect humanity would be that in which these two forms of conscious activity should attain their full development' (*CE*, p. 281). As he maintains, the development of instinct, which gives rise to intuition, was lost when animals gave way to humans. Bergson also admits that there may have been times and places where evolution produced sentient beings which we would call men but 'who are not necessarily our ancestors' (*TSMR*, p. 273n).

4   Instinct and Intelligence

Bergson distinguishes two culminating points of evolution resulting from two separate lines: the human species at the top of the vertebrates and hymenoptera at the top of the arthropods. In hymenoptera the instinct, and in man the intelligence, have reached their highest level. Instinct and intelligence are, according to Bergson, two tendencies, 'two modes of acting on the material world' that the force of life has to choose from (*CE*, p. 149). It can affect the material world directly, creating an organized instrument to work with, or indirectly by making an organism construct an instrument out of inorganic matter. The reason for intelligence being prevalent in humans is the lack of ready-made organized instruments to sustain their existence.

Instinct is a natural ability to use an inborn mechanism, a continuation of the work of organization itself, for it is not clear when the activity of nature gives way to the activity of instinct. 'We may say, as we will, either that instinct organizes the instruments it is about to use, or that the process of organization is continued in the instinct that has to use the organ' (*CE*, p. 147). So, in their perfected forms, instinct is a faculty of using and constructing organized instruments (organs) and intelligence is the faculty of making and using unorganized instruments (manufacturing tools). An instrument that instinct uses is self-repairing, complex in its structure, simple in its function and perfect in its action. The drawback there is that its structure is invariable, because a change in it would signify a modification of the species. Therefore instinct is highly specialized, 'being nothing but the utilisation of a specific instrument for a specific object' (*CE*, p. 148).

As for the instrument constructed and used by intelligence, it is imperfect but flexible in its form: it can be modified and used for any purpose. Also, unlike the instrument of instinct, it is open to change and improvement and offers more and more freedom to its user to achieve a higher level of constructive machinery. Bergson thinks that if the creative force of life were unlimited, it might have developed intelligence and instinct in the same species, but as it is limited and must choose a direction, instinct and intelligence became separated.

In many cases instinct is unconscious, Bergson finds. But there is a difference between unconsciousness in which consciousness is absent and unconsciousness in which consciousness is nullified, i.e. impossible in principle. In certain cases when the consciousness is absent, as in habitual actions, the act fits its representation so perfectly that there is no room for consciousness there. On the other hand, when the act does not follow the representation adequately, we have consciousness of the gap. Thus consciousness appears in cases of possible actions and indicates hesitation or choice. And, 'where the action performed is the only action possible', consciousness is superfluous and is reduced to nothing (*CE*, p. 152).

Consciousness is now being defined as 'an arithmetical difference between potential and real activity. It measures the interval between representation and action' (*CE*, p. 152). Seen in this light, consciousness should be expected to accompany intelligence and be absent in the instinctive activity. However, Bergson asserts that both intelligence and instinct involve knowledge, reflected upon and conscious in the former, and acted and unconscious in the latter.

At this stage the story twists as we learn that Bergson believes both instinct and intelligence to be innate: a baby would instinctively look for its mother's breast, and also, as a toddler, immediately understand relations between things such as cause and effect, like with like, content to container etc., which, to Bergson, demonstrates intelligence. This is interesting, because in *Matter and Memory* he had explicitly argued against any possibility of images being produced by the brain, the argument that he used as a necessary foundation for the claim that spirit (or mind, or the inextensive) is not derived from the material component of the brain but is correlative with it (*MM*, p. 9). We could have interpreted this part of Bergson's theory as a campaign against innate knowledge. What *Creative Evolution* seems to tell us now is that, even though Bergson denied that any factual knowledge could be innate and coming from the brain, the predisposition to receive data as knowledge about the environment is innate. In fact, this seems to be merely a rephrased Kantianism.

By allowing innate knowledge as a predisposition for factual knowledge, Bergson is demonstrating that the stream of conscious or quasi-conscious activity, which passes from an individual to its siblings, is not interrupted and then restarted in another individual, but that its continuity is preserved even though it is modified in such a way that each individual has an impression that his or her conscious autonomy is complete and that his or her knowledge and memories are entirely his or her own. This is an illusion, as the individual's ability to know has been inherited from the infinite chain of other individuals, a fact that lowers the degree of one's own autonomy.

Firstly, we have knowledge of things, and secondly, of relations, Bergson asserts. Innate knowledge of things, matter in its immediacy, is instinct, whereas innate intelligence is knowledge of a form, i.e. of the totality of all relations. In instinct, we have knowledge applicable to one specific object; in intelligence, we have knowledge containing form with no matter and applicable to an infinite range of objects.

The above is an analysis of instinct and intelligence from the standpoint of knowledge rather than that of action. But Bergson is leading us to the unification of ontology and epistemology, of being on the one hand and creating and knowing, on the other. At this stage Bergson says explicitly that knowledge and action are two aspects of the same faculty.

Instinct is innate knowledge of a thing because it is a faculty that uses a natural instrument, the philosopher continues, and it must have knowledge of this instrument and of the object of its application. Intelligence is a faculty that constructs an artificial instrument and, therefore, an ability to find the way out of a difficult situation, and it is for that purpose that intelligence is primarily the tendency to establish relations. This innate formal knowledge can be filled with any content and is not restricted to any content in particular. Thus 'an intelligent being bears within himself the means to transcend his own nature' (*CE*, p. 159). However, the purely formal character of intelligence lacks the ballast enabling it to concentrate on the object. Instinct has the desired materiality, but it cannot speculate. Hence Bergson's conclusion: 'There are things that intelligence alone is able to seek, but which, by itself, it will never find. These things instinct alone can find; but it will never seek them' (*CE*, p. 159).

Refusing to grant intellect the too high a role of an absolute that knowledge depends upon, Bergson regards it 'as relative to the needs of action' (*CE*, 161). And this is another argument he uses to merge ontology and epistemology, as knowledge then is no longer a product of the intellect but is part of the ever-moving reality.

Intellect constructs instruments out of inert solid matter, and even if any organized or fluid material is involved in the process, it is treated as inert and solid, the fluidity and the living escaping the attention of intellect. '*Our intelligence, as it leaves the hands of nature, has for its chief object the unorganized solid*' (*CE*, 162). It presents everything real as consisting of parts, units, thus creating arbitrarily a discontinuity out of the real, for '*of the discontinuity alone does the intellect form a clear idea*' (*CE*, 163). Also, when it comes to mobile objects, we are not interested in the process of their movement, but predominantly in the initial and consecutive positions of the bodies involved.

Movement is the reality itself, and immobility is secondary to movement, according to Bergson, but our intellect, practically orientated, ignores movement, concentrates on immobilities and then reconstructs movement out of immobilities. The result of this operation is not movement as it really is but its practical equivalent – the spatial image that we have of movement, its trajectory.

When manufacturing tools our mind looks at matter with a view to modifying it and thus regards it as a substance indifferent to any form and capable of adopting any form. There is a limit to what extent real matter can be decomposed and reassembled, but the mind disregards the matter's real limitations and, in principle, treats it as decomposable in the homogeneous space that underlies it and this forms the plan of our potential actions on things. So, 'the intellect is characterized by the unlimited power of decomposing according to any law and of recomposing into any system' (*CE*, 165).

The aforementioned deficiency of the intellect is reflected in language. By the analogy with spatial objects, we treat concepts as if they exist outside each other and as if they are, like objects, stable and solid. 'The intellect represents *becoming* as a series of *states*, each of which is homogeneous with itself and consequently does not change' (*CE*, 171). If we attempt to reveal the internal change occurring within one of those states, we break it up into another series of states, and so on, Bergson explains, and thus becoming escapes our understanding. What is new escapes it also because we are always trying to reconstitute what is already given and reject the unforeseeable and so we reject creation. Causality, that we find everywhere, accommodates repeatable causes, old and known, and repeatable effects, and does not allow us to see unique causes and unique effects of creation. Just like becoming, novelty, which is an essential aspect of life, escapes us. Ultimately, Bergson says that the intellect is not an instrument designed to understand the mobile and living: '*The intellect is characterized by a natural inability to comprehend life*' (*CE*, p. 174). We can read into this that intellect is equally unable to comprehend duration, since, like life, duration is movement and becoming.

The faculty that can comprehend life is instinct. It is moulded from the very form of life and, while intellect employs a mechanical approach to everything, instinct proceeds organically. Instinct is a prolongation of the work of life and, if it were conscious, 'would give up to us the most intimate secrets of life' (*CE*, p. 174). Unfortunately, what is instinctive cannot be expressed in terms of intelligence.

Bergson attempts to explain instinct not in terms of intelligence but declaring it to be sympathy (*CE*, p. 186). The term 'intuition' appears in *Creative Evolution* meaning 'instinct that has become disinterested, self-conscious, capable of reflecting upon its object and of enlarging it indefinitely' (*CE*, p. 186). Man's aesthetic faculty shows that such an effect is not impossible as, according to Bergson, an artist creates a work of art by a simple move which is not unlike the creation of organs by nature.

Bergson envisages a possibility of intuition compensating for the deficiency of intelligence and transcending it. The two faculties of knowledge, intellect and intuition, would complement each other, he believes, because without the push from the intelligence, instinct could never become intuition, and without intuition, intelligence can only restrict itself to a mechanistic view of life. In Bergson the double form of consciousness corresponds to the double form of metaphysics,: intelligence corresponds to inert matter, and intuition, to life.

To summarize: intelligence is manifested as manufacturing ability, materialized invention, whereas instinct is manifested as a natural ability to use an inborn mechanism, continuing the work of nature. Intelligence is described as a faculty of making and using unorganized instruments, and instinct is a faculty of using and constructing organized instruments.[8] If intelligence allows infinite variations, instinct is highly specialized. Intelligence involves thought and consciousness, and instinct is unconscious in the sense that consciousness is absent from instinct.

H. Wildon Carr, albeit sceptical about treating instinct as a form of psychological cognition at all,[9] points out that consciousness and uncon-

---

8   The evolutionary conception of humans as essentially tool-making animals is criticized and debated by philosophers of technology. An illuminating collection of essays on this subject can be found in Robert C. Scharff, Val Dusek, eds, *Philosophy of Technology* (Oxford: Blackwell Publishing, 2002).

9   H. Wildon Carr, 'Bergson's Theory of Instinct', *Proceedings of the Aristotelian Society*, Vol. 10 (1910), 110–12.

sciousness are not main characteristics differentiating between instinct and intelligence as Bergson allows degrees of consciousness in both faculties.[10] Kolakowski, thinking along the same lines, makes an attempt to reconcile intuition and analysis in Bergson by softening the opposition between them. He points out that, although Bergson portrays them as radically different ways of being acquainted with an object, they primarily differ in the object of their application rather than in the method: intuition, the critic finds, is limited to approaching life, mind and movement, and everything else must be dealt with by analysis.[11] Therefore, 'the intuition/analysis dichotomy ... is ... less sharp than it might appear from his [Bergson's] general definition'.[12] Analysis and intuition thus complement each other, and neither can replace the other, since they differ in the area of their application, Kolakowski concludes. This, of course, can be challenged by recalling that for Bergson, everything ultimately is duration and movement; so to say that intuition is limited to grasping movement would be the same as saying that intuition is limited to grasping being. Kolakowski himself inevitably admits the universality of possible application of intuition. For Bergson, he remarks, intuition is 'an act of identification with the time in which the "object" is immersed'.[13] Admitting that intuition grasps the temporality of the real here is the same as saying that intuition grasps the object, since, for Bergson, the temporality of the object is identical with everything that the object entails.

Perhaps Bergson's own words can determine the intensity of the opposition between intellect and instinct. In his letter to H. Wildon Carr,

10   Carr, 'Bergson's Theory of Instinct', 101.
11   Moore reads Bergson differently here: 'What is new is Bergson's willingness to make the contrast between intuition and analysis entirely general (*independent of subject-matter*)' (italics mine) (Moore, *Bergson: Thinking Backwards*, 8). Both Moore's and Kolakowski's reading of Bergson are legitimate due to the obvious inconsistencies in Bergson's own texts. However, when one realizes that the ultimate genuine object of knowledge is temporal reality, then the ultimate genuine cognition is only possible in intuition, and analysis only gives us an illusionary idea of illusionary reality.
12   Leszek Kolakowski, *Bergson* (Oxford: Oxford University Press, 1985), 32.
13   Kolakowski, *Bergson*, 35.

Bergson emphasizes that in his understanding of the problem of cognition, intelligence is surrounded by a fringe of intuition, which allows us to sympathize with the essential features of life. He says that one can call this fringe 'intelligence', albeit stretching the meaning of the word, and that in his view, this fringe is more like instinct than intelligence.[14] So Kolakowski is probably correct in softening the Bergsonian opposition between the supposedly incompatible and conflicting intelligence and instinct. Also Cunningham observes that, starting with the sharp opposition, Bergson resolves it by indicating that ultimately, analytical and intuitive knowledge do not contradict but complement each other.[15]

5   Matter and Consciousness in Evolution

In Bergson's theory of evolution, the assumption that matter and consciousness originated from the same source grows into another assumption that along the way of their development, matter and consciousness have been shaping each other. Next, this idea of the mutual adaptation of matter and consciousness is taken into the domain of action. Bergson believes that intellectuality and materiality must have been developed alongside each other and by reciprocal adaptation, both being 'derived from a wider and higher form of existence' (*CE*, p. 197). If Bergson is right in claiming that matter and intellect have been involved in a process of mutual adaptation, there is no reason to suspect that this process has stopped. If the primary purpose of intellect, as Bergson asserts, is to act on matter, then we should turn our attention to action.

14   Carr, 'Bergson's Theory of Knowledge', 60.
15   Gustavus Watts Cunningham, *A Study in the Philosophy of Bergson* (New York: Longmans, Green and Co., 1916), 39. The detailed analysis of the dual position that Bergson holds in respect of this opposition see Chapter 3 in this book, 32–64.

Looking at the relation between consciousness and matter from the standpoint of Bergson's theory of action, they appear as parts of one and the same continuity, and as a cause and effect reciprocally. Material circumstances question consciousness and make it produce a response; consciousness initiates action, a process in which consciousness and matter are interwoven with each other, followed by modified matter existing independently from consciousness. This modified matter in its turn will act as a cause for a consciousness to act in response to it. It appears as if in this type of fluctuating cyclical relation, consciousness and matter move from being distant from each other to merging, with the culminating point being at the time of action, and then distancing from each other again.

It is true that if we take consciousness and matter as beings of a different nature and originating from different sources, then the relation between them will remain inexplicable; for how do we find a mediating agent that ensures co-operation of mind and body, or how can we explain any contact between consciousness and the material world at all? This is exactly what was lacking in *Matter and Memory*. The transition from intelligence to physicality, or from physicality to intelligence, appears as an inexplicable qualitative leap.

However, if we agree with Bergson that 'the intellectuality of mind and the materiality of things' are created by the same movement, and that the relation between intelligence and matter is that of mutual adaptation which is the inversion of that movement (*CE*, p. 217), then at least there is a starting point on the way to try to understand the connection between consciousness and matter.

In living bodies, matter and consciousness are part and parcel of the same thing – a living process, with different degrees of consciousness being present in all life forms. The point of origin of a concrete action is found in the brain, and that is how Bergson describes their relation.

> Everything seems ... to happen *as if* consciousness sprang from the brain, and *as if* the detail of conscious activity were modelled on that of the cerebral activity. In reality, consciousness does not spring from the brain; but brain and consciousness correspond because equally they measure, the one by the complexity of its structure and the other by the intensity of its awareness, the quantity of *choice* that the living being has at its disposal. (*CE*, pp. 276–7)

Further on he says:

> The consciousness of a living being ... is inseparable from its brain in the sense in which a sharp knife is inseparable from its edge: the brain is the sharp edge by which consciousness cuts into the compact tissue of events, but the brain is no more coextensive with consciousness than the edge is with the knife. (*CE*, p. 277)

Intelligent activity involves both brain and consciousness as inseparable elements of the same process. In action, involving body as a mediating agent between consciousness and the world, the relation between the latter two becomes complicated by the fact that intelligence is not the only form of consciousness that is involved in the process. Pure consciousness alone acting via brain is capable of reflecting and making decisions, but it is not capable of physical contact with the outer world. The body that does the conscious acting comprises a multitude of organs, which are composed of a multitude of cells, each of which, according to Bergson, is like a living organism in itself, and as such contains consciousness in the form of instinct which ensures the organic existence of the living.

Between consciousness that correlates with the brain, and matter that is acted upon, we have a multitude of conscious processes that correlate with the body. The further we move away from the consciousness of the brain, the more we find that the processes lying in between are loaded more with matter than with consciousness. Thus in action or, more specifically, in manufacturing, we observe a gradual descent from the ultimate intensity of awareness in the brain to the ultimate passivity of the manufactured object whose relation to consciousness is one of a receptacle of an imprint.

The hierarchy of the universal 'consciousness – matter' movement, emerging in *Creative Evolution*, seems to be as follows:

1 At the highest point there are brain and consciousness where the intensity of conscious awareness is at its strongest.
2 One step down there is a body which is largely material, and yet its materiality is loaded with consciousness of the living process, which presents itself as instinct. Also, the body is an active conductor of the conscious will.

3 Further down we find tools that the body uses to create artefacts.
These tools are material things but they serve to prolong the work of
consciousness – they are passive conductors of consciousness.

4 At the bottom there are artefacts which are matter that has received
an imprint of consciousness.[16]

5 Beyond this point there is matter untouched by consciousness,
indifferent to it and independent from it.

6   Human Consciousness Prolonging the Evolution

Bergson talks of the movement of consciousness running through the evolutionary movement, from the starting point to the highest known point,
the human intelligence, where it is intensified and where it culminates.
Also, whereas from the beginning of life to human species consciousness
followed the direction of ascent through the living, in the human mind
it is *refracted* and its further movement as conscious activity and, more
specifically, manufacturing, is descent into matter.

In line with Bergson's thought we can observe that we live in a world
saturated with consciousness via two major movements: the movement of
ascent that involves the whole of organic life with human consciousness
at its highest point, and the movement of descent, with its multitude of

---

16   The issue would be immediately complicated if we questioned Bergson's theory of
action against the distinction of more and less basic actions. According to this distinction, one does one thing by doing another, such as voting by raising one's arm,
whereby the raising of the arm is more basic than voting. One would probably have
to admit a regress in the conscious component within one and the same physical
act, and introduce layers of reality with different intensity and extensity ratio, where
raising the arm or, in the case of manufacturing, shaping clay would be more material
than conscious, and voting, or making a mug, more conscious than material. (For
the theory of basic actions, see Jennifer Hornsby, *Actions* (London: Routledge &
Kegan Paul, 1980), Chapters 5 and 6.).

specific activities, and without that unifying force of evolution wherein each individual consciousness acts as if it is independent and detached. If in the ascending movement, consciousness is more instinct than intelligence, and is ontologically bound with material living bodies, in intelligence, consciousness is much freer and its ontological bond with the material base is weaker as it is able to choose an object of its attention with a greater degree of freedom than instinct. Brought to its highest point by evolution, human consciousness begins to act as if it were the primary cause of the changes occurring through man's activity, as if the evolutionary development leading to man has little to do with what each individual consciousness chooses to change. And if in the ascending movement we can talk of the unity of all processes, the descending movement is broken up into a multitude of dissipated movements of independent streams of consciousness. It is in this sense that we understand Bergson's assertion that 'man ... continues the vital movements indefinitely, although he does not draw along with him all that life carries in itself' (*CE*, p. 280).

As we try to show in Diagram 4, in manufacturing consciousness is realized through action and leaves its imprint on an artefact. It remains totally external to the artefact; the specific shape, look, smell and colour of the artefact are like footprints left by consciousness. And, although consciousness remains external to the manmade object, it is *there*, and it is recognisable.

We can add that it matters greatly to archaeologists whether the discovered stone acquired its specific shape naturally, or whether its natural shape has been altered deliberately by the first man who made it into the first tool. Equally, travellers passing through uninhabited land would be excited to find any man-made objects – the remains of a tent, or empty cans – and an immediate feeling of having some communication with human beings.

However, consciousness carved into man-made objects remains superfluous to the matter on which it is inscribed. Consciousness springs off the artefact only towards another human being and remains dormant when no one is there. Objects themselves remain fragments of matter and do not partake of consciousness. This, we can assume, stands as an obstacle in the way of manufacturing artificial intelligence: a most sophisticated machine designed to speak, calculate and move about would not acknowledge these processes internally, and would remain a collection of material particles indifferent to the external attempts to inject consciousness into it. When a material body gives in to the conscious effort coming from outside, it gives in to the physical force, unaware of the consciousness that stands behind it.

Man:
the highest form of conscious activity

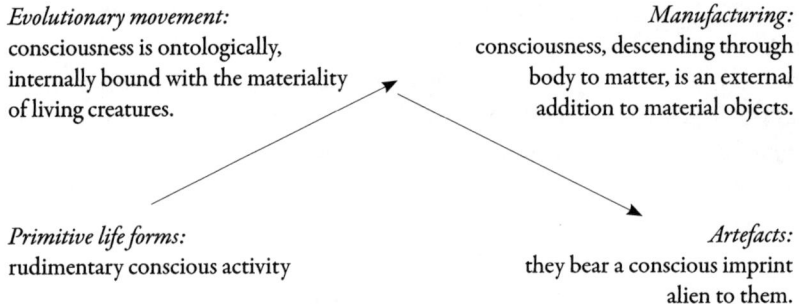

Evolutionary movement:	*Manufacturing:*
consciousness is ontologically,	consciousness, descending through
internally bound with the materiality	body to matter, is an external
of living creatures.	addition to material objects.

Primitive life forms:	*Artefacts:*
rudimentary conscious activity	they bear a conscious imprint
	alien to them.

Diagram 4 Consciousness in the world

In organic bodies consciousness is an integral part of their being and a driving force of their form of existence which is internally connected with their creation wherein an act of will (or quasi-will) creates directly. In material artefacts consciousness too, is the driving force of their creation out of raw matter, but here consciousness cannot act directly – it must use mediating elements such as body and tools. Also, consciousness always remains outside the object and can never become an integral part of its being. Consciousness in the artefact can be recognized only by another consciousness, but not by the object itself.[17]

---

17   Any further pursuits to follow up conscious input into material artefacts would need to refer to Ilyenkov's objective idealism. According to Ilyenkov, manufacturing involves giving a new physical form to the material object (pieces of wood) as well as attributing to the object certain significance (table), which has nothing to do with the physical substance of the object. This significance is the objectively existing

The hierarchy of movement in the living process, as it appears in Bergson's theory, can be summarized in the following terms. The work of an organ is restricted to a particular spatio-temporal point. Instinct is a prolongation of the work of an organ. Introducing extra mobility, it breaks the spatial immediacy of the body but, unable to be aware of anything but the immediacy, it is confined to the temporal immediacy which it prolongs into the next moment. It is true that the end of the work of instinct may lie far in the distance (reproductive behaviour, for example), but this end at a distance is not known through instinct, and instinct appears concerned only with the prolongation of the here-and-now into the next moment. Intelligence goes beyond maintaining the continuity of life and satisfying natural requirements and concerns itself with the future, ensuring that the continuity of life is maintained in the remote future and not just in the present and in the immediate future.

As for the place that an individual self occupies in the hierarchy of being, we have learnt that firstly, the duration of selfhood is modelled as a living process, and is itself a fragment of that living process, enriched by intelligence. In the structure of duration, there are various currents: the current that is but a prolongation of the stream of life, manifesting itself as innate cognitive abilities common to all humans; the current that is a prolongation of concrete hereditary features such as traits of character; the concrete developed intelligence and concrete traces of instinct that are addressed to concrete content and are responsible for the self-awareness of the individual, that part of the individual that defies the evolutionary movement and regards its own being as the end of all prior evolutionary development.

'ideal form' of the object. See E. V. Ilyenkov, 'The Concept of the Ideal', *Philosophy in the USSR: Problems of Dialectical Materialism* (Moscow: Progress, 1977). For an interpretation by a Western commentator, see David Bakhurst, 'Lessons from Ilyenkov', *The Communication Review*, Vol. 1(2) (1995), 155–78.

CHAPTER 5

Heterogeneous Duration

1   Heterogeneous Duration: Non-Numerical Multiplicity

The brief exposition of the three major Bergsonian texts (see Chapters 2–4) demonstrates that his philosophy points towards a theory of the universal principle of being – heterogeneous duration. However, this theory is incomplete because the discussion of duration is fragmented and its exemplification as concrete modes of being is not systematic. We will attempt to develop a plausible version of heterogeneous duration, complying with Bergson's key principles where possible, and then apply it to concrete reality. This chapter will be centred on the general metaphysical aspects of duration.

The idea of heterogeneity is necessary to Bergson's project of making a radical distinction between simultaneous and successive phenomena, which form the continuity of identity of a temporally stretched unity.[1] The key difference between a spatially stretched unity with all its elements being simultaneous and a temporal unity is that, according to Bergson, the spatial unity can be divided into parts with clear spatial borders, whereas the temporal unity cannot be broken down into segments which can be regarded as separate from one another. Thus, he claims in *Time and Free Will*, the temporal continuity is indivisible. In *Matter and Memory* we learn that ultimately, spatial reality is also indivisible (*MM*, p. 196), because

---

[1]   Čapek offers a thorough analysis of the Bergsonian heterogeneity with links to other key notions of the Bergsonian philosophy. See Čapek, *Bergson and Modern Physics*, 83–186, esp. 118–25; 142–51. A brief explanation of the Bergsonian heterogeneity can be found in Lacey, *Bergson*, 49–50. Moore's account of heterogeneous properties of duration is also useful: see Moore, *Bergson: Thinking Backwards*, 54–65.

after all, all reality is temporal, even that which seems purely spatial. Thus Bergson opposes the idea of temporal duration not to space but to our illusory idea of purely spatial reality.

For Bergson, if elements of continuity cannot be separated, this means that they cannot be treated as its countable subunits. On the other hand, the content of temporal continuity is not the same throughout its development; it differs from its previous stages, firstly because every event evolves, and secondly because it simply becomes older and accumulates the time of its own swelling existence. This makes Bergson assert that in its essence, duration is a non-numerical qualitative multiplicity. But this is not all; he also talks of the elements of duration permeating one another. He uses metaphors of a ripening fruit and a snowball in order to show that the past stages of duration do not disappear but become integrated into the state in which the duration finds itself at present.

The example of a musical piece is supposed to provide a decisive demonstration of a phenomenon which is incomprehensible in any other way than as an unbroken temporal unity with elements diverse but inseparable from each other, and whose meaning depends on the other elements. To spell this out, let us imagine a musical phrase *abcda*. All these notes can be taken as separate units. They can be played in isolation, or in a different order. Each of them can also be combined with other notes in another musical phrase. For instance, the musician can single out the note *d*, play it alone, or in a sequence *cgad*. But as a member of the phrase *abcda*, the identity of the note *d* cannot be defined solely by its physical immediacy. The musical meaning of *d* is defined by its relations with *abc* and *a*. Isolated from *abc* and *a*, alone, or as part of a phrase *cgad*, the meaning and identity of *d* will be different. Thus the content of *d* spills over out of its own physical and temporal immediacy and spreads beyond itself involving the surrounding elements.

An attempt to enumerate qualities in such heterogeneity would mean treating *a*, *b*, *c*, *d* and *a* as units which represent their content by themselves. But such isolation kills the music. As we can see, since each note contributes to the content and identity of each other note, then enumeration of them as units becomes very complicated. If we wanted to treat *d* as a countable unit, we would have to assume by this that *d* exhausts its own content and is complete without reference to *abc* and *a*, and also to *abcda*. But to complete *d*, we must involve other notes in the phrase as well as the complete phrase itself. That defies the idea of presenting *abcda* as a collection of countable units, and *d* as one of those units. So, *abcda* must be accepted as a unity which entails multiplicity, as a heterogeneity in the Bergsonian sense. Or, as Moore usefully observed, it may be said that heterogeneity entails complexity rather than multiplicity.[2]

The above exemplifies a heterogeneity of an obviously temporal piece of reality: a musical piece needs time to be brought to life. But the idea of heterogeneity is supposed to be applicable to the reality of everything existent, which is necessarily, albeit not always obviously, temporal. Let us test another multiplicity of qualities: we drink tea, which is liquid, hot and sweet. Could we treat all three qualities as separate units? Sweetness could be such a unity, if we could isolate it and point at it ostensibly. We taste the sweetness, but it is inseparable from the liquid and the heat. Sweetness cannot appear by itself; wherever sweetness occurs, it needs to be accompanied by other properties. It does not have to be tied with liquid and heat specifically: if it is the sweetness of an apple, it will be tied to the hardness and coldness of the apple's texture. But whatever its accompanying properties are, it cannot be felt in separation from them, nor can it be separately located. In either tea or apple, we have a heterogeneity where the identity and content of every element depend on everything else.

In heterogeneity the whole obviously consists of diverse components, but they are not extractable as individual items with definite outlines. But how do we know that they remain distinct qualities, have not become homogenized and maintain the heterogeneity of the whole? Unexpectedly, help comes from the remote philosophical past, from Thomas Aquinas' *Summa Theologica*. In *Treatise on the Angels*, he faces a problem, similar to ours. He needs to account for the multiplicity of angels, linking it with his earlier conclusion that the latter are not corporeal.[3]

---

2   Moore, *Bergson: Thinking Backwards*, 64.
3   Thomas Aquinas, 'Treatise on the Angels', in Thomas Aquinas, *The 'Summa Theologica' of St Thomas Aquinas* (New York: T. Baker, 1911), 297.

Aquinas' position is that the form explains the species membership, and matter explains the individuality of things, but this does not work for angels because they are not material. Since angels do not contain matter and form, they cannot be distinguished as separate individuated units and cannot be enumerated. Yet, Aquinas needs to demonstrate the diversity of angels, as he would not want to assert that there is just one angel, because it is only God that is one. Aquinas' solution is that angels differ as do species or kinds, similar to qualities within the Bergsonian qualitative multiplicity. There is also the possibility of explaining matters further, outlined by Aquinas but not pursued by him. Aquinas wants to prove that angels cannot belong to one species, and one of his passages reads as follows. 'And if the angels had matter, not even then could there be several angels of one species. For it would be necessary for matter to be the principle of distinction of one from the other, not, indeed, according to the division of quantity, since they are incorporeal, but *according to the diversity of their powers*'.[4]

Matter cannot explain the diversity of powers, according to Aquinas, but we will say that the diversity of the powers of the events in itself explains differentiation. What Aquinas provides here, in effect, is identification of an entity, which would be an alternative to the ostensive visual singling out of it. An entity, we read in this, may be identified as such according to the effect it has on the rest of the world, the effect which would differ from effects produced by other entities. For the purpose of this enquiry, we can apply Aquinas' remark on angels to reality in general and distinguish qualities within a heterogeneous multiplicity by *relations* that they emanate, if the direct spatial or quasi-spatial identification is not possible.

The key feature of a qualitative non-numerical multiplicity must be the interrelatedness of all qualities involved in its construction. For if there are distinct qualities, they must produce distinct relations with each other, or else the difference between the qualities would fade away, and the heterogeneity, the multiplicity of qualities, would collapse into homogeneity, and become just one quality. In other words, each quality will have to demonstrate the way in which it remains opposed to other qualities, albeit

4   Ibid., italics mine.

escaping identification as a phenomenon occupying a spatio-temporal position, reserved for that phenomenon only.

Returning to our example of the heterogeneity of music, we could say that *d* is a necessarily qualitative component of *abcda*. When we listen to *a*, *b*, *c*, *d* and *a*, we bear in mind the entire sequence, but the effect that each element produces in us differs. The first *a* may evoke a sense of anticipation and a desire to hear the whole phrase; *b* and *c* could represent a build up of emotional tension; *d* would be its culmination and the second *a* would give us a sense of anticlimax and bring the phrase to a conclusion, as if it had been a mini journey, where we started out at *a* and returned to *a*, enriched by the experience of *b*, *c* and *d*. All the notes, albeit entwined with each other, produce a different effect on the listener and in this they remain distinct from each other. As for the liquid, heat and sweetness of tea, they too produce different effects on us even though they cannot exist or appear in separation. In fact, it is their effects in the first place that make us distinguish them as individual qualities.

2   Constructive Negation: Same and Other

Bergson explicitly argued in *Creative Evolution* that duration which equals being does not involve negation, because negation is not an ontological but an epistemological feature.[5] We will argue that although Bergson is opposed to negation understood as a void, the idea of duration does not exclude negation if it is understood in terms of sameness and otherness. For if we follow Bergson and cancel negation altogether, then the theory of duration can be taken as a one-sided affirmation of fullness – as Bachelard understands it.

The very idea of temporal heterogeneity, Bachelard observes, with the diversity of its qualities taking place over a period of time, means the presence of some qualities and the absence of others, and when absence

5   See Chapter 4, Section 1 for an exposition of Bergson's theory of the nought.

gives way to presence, we have a change from negation to affirmation, and that in itself is a leap. Bachelard claims that in order to accept continuity strictly in the Bergsonian sense, we would have to give up heterogeneity and instead assert the homogenized content of duration, which is the same at all times. Following this, the idea of novelty and becoming would have to be given up too, as absolute sameness of content would not allow for novelty and change. Of the living duration Bachelard says: 'So great is heterogeneity of its terms that succession is in effect discontinuity'.[6] Bachelard's ambition is to improve the Bergsonian continuity by bringing discontinuity into it: 'Time is ... continuous as possibility ... It is discontinuous as being'.[7] He introduces lacunae into duration, trying to reflect the 'duality of events and intervals'.[8]

Describing Bergson's philosophy as 'a philosophy of fullness',[9] Bachelard says that there is no room in it for contradiction or emptiness of any kind.

To counteract continuity which can possibly be understood as the sameness of the temporal content of duration, Bachelard introduces into it nothingness. Bergson explicitly excluded the nought from ontology claiming that negation of x is secondary to affirmation of x,[10] and Bachelard is sceptical about this. Convinced that the dialectic of being and non-being is the foundation of ontology, he links it to the idea of becoming preceded by a void, for as he observes, 'the miracle of being is as extraordinary as that of resurrection'.[11]

Arguing against the one-sidedness of Bergson's argument that the idea of nought is parasitic on the idea of being, Bachelard remarks, '[W]hile it is very true that you can only empty what you first found full, it is just as accurate to say that you can only fill what you first found empty. ... [T]here seems to us to be a perfect correlation between emptiness and fullness. One is not clear without the other and in particular, one idea is

6   Bachelard, *The Dialectic of Duration*, 42.
7   Bachelard, *The Dialectic of Duration*, 44.
8   Bachelard, *The Dialectic of Duration*, 19.
9   Bachelard, *The Dialectic of Duration*, 23.
10   See Chapter 4, Section 1.
11   Bachelard, *The Dialectic of Duration*, 32.

not clarified without the other'.[12] In other words, Bachelard finds negation to be a necessary background for affirmation.

It can be said that Bachelard's efforts in this respect are unnecessary because Bergson's heterogeneity already accounts for discontinuity via the notion of a continuingly growing multiplicity of diverse qualities. Bergson himself admits of leaps in duration, when he recognizes that the evolutionary movement is not a smooth process. 'But it proceeds rather like a shell, which suddenly bursts into fragments, which fragments, being themselves shells, burst in their turn into fragments destined to burst again, and so on for a time incommensurably long' (*CE*, p. 103). As Marie Cariou observes, 'Bachelard raisonne comme si la durée était une substance homogène. Mais c'est bien pour éviter au contraire cette interpretation que Bergson parle d'élan, de directions, de tendances'.[13] She points out that Bachelard's erroneous perception of duration as a smooth continuity is due to an overestimation of Bergson's comparison of duration with a melody. She reminds the reader that Bergson also talks of turbulent currents and of swelling rivers. This diversity of images, she suggests, is supposed to illustrate the diversity of qualities of duration, which should not be restricted to one definition based on only one image. The Bergsonian duration, according to her, involves efforts and overcomes obstacles, which contradicts the image of a quietly flowing river.[14] I would add that Bachelard's misinterpretation of duration could have been avoided if Bergson had not argued against negation in the way he did. Accepting his claim that one must not posit the void prior to positing duration, we will reintegrate the nought into duration in a different role.

It seems that the difficulty in the way of understanding the nought is that philosophers, including Bergson and Bachelard, expect it to be a quality, polar to being but equal to the latter as an ontological category.

12   Bachelard, *The Dialectic of Duration*, 30.
13   'Bachelard treats duration as if it were homogeneous substance. But this is so as to avoid that interpretation of duration where Bergson talks of drive, of directions, of tendencies'. (Marie Cariou, *Bergson et Bachelard* (Paris: Presses Universitaires de France, 1995), 70). (Translation mine.)
14   Cariou, *Bergson et Bachelard*, 70–1.

Any suggestion to consider reality in terms of the dialectic of being and nothingness equates the roles of being and the nought, where being actively creates and the nought actively destroys. As a quality, the nought must be able to sustain itself and be comprehensible as a self-contained phenomenon. However attempts to grasp the nought without any reference to being collapse into absurdity because the nought then should be responsible for maintaining its own existence, its own being. As soon as one realizes that, one will inevitably say that the idea of the being of the nought contradicts the idea of the nought itself, because it is supposed to be absolute non-being. That is why Bergson is forced to conclude that the nought and negation in general are artificial constructs, secondary to the reality of absolute affirmation, and then he excludes negation from ontology.[15]

We are prepared to talk of being and nothingness as parts of a heterogeneous whole, but with a necessary asymmetry between these terms. We will not seek in nothingness qualitative features and the ability to sustain itself as an absolute void and, contrary to Bergson, we will include it in the objective world. We will say that concrete manifestations of being within themselves contain non-being of some properties and themselves radiate their own non-being in other places and at other times. In this, we are taking on board Plato's position developed in *Sophist*, where not-being of some properties is taken as the being of other properties rather than an absolute void.[16] The difference between the Platonic position and ours, Bergsonian, will be that we will add a dynamic dimension to being and not-being. If Plato merely states the being of some property and points out its not-being replaced by the being of another property, we are saying that the being of a property *produces, generates* its own not-being beyond its own boundaries.

Whereas being in its self-evident self-affirmation is a quality, negation of its presence at other times and other places is a relation. As a quality, each embodied manifestation of being emanates a halo of numerous relations, and amongst them there is *not having* certain properties, or *not being* in any other places rather than in a particular one.

---

15   For a recent discussion on nothingness, which brings together Western and Japanese philosophers, see James W. Heisig, *Philosophers of Nothingness* (Honolulu: University of Hawaii Press, 2002).

16   Plato, *Sophist* (Whitefish, Montana: Kessinger Publishing, 2004), 23–5.

Negation taken in this sense is ontologically secondary to being understood in qualitative terms, albeit it does not mean, of course, that it is temporally later than being. It is a relation which is not made up by a rational observer but is ontologically presupposed. The facts that a given table is not white and is not in this room are not mere 'pedagogical' constructs, as Bergson would put it, because for the table to be described as not white and not in this room, it must *not* be white and *not* be in this room in its actual reality.

As part of reality, negation appears in the following guises. It can be the non-existence of some quality at time t, possibly followed by the existence of that quality at time t'. Or it can be that an entity displays some qualities and *not* others. One may argue that these relations require a perceiver in order to be ascribed to a quality, and that therefore they do not, strictly speaking, belong to the quality. From the Bergsonian standpoint we will say that even if the perceiver may be required to acknowledge a relation, the acknowledgment is itself a relation, and it takes two terms to give rise to a relation. So the relation of acknowledgment cannot be created unilaterally by the perceiver. If it could, then it would be up to the perceiver to ascribe any relation to any quality, but this is not so. Relations, even those articulated by the perceiver, are regulated by qualities themselves.

The negation as a relation is not a mighty rival of being, equal to being and posing as a monster that gnaws on being: this negation has a constructive role to play in the formation of heterogeneity. For instance, we can distinguish between duration at time t_1, (d_1), and the same duration at time t_2, (d_2). D_2 is richer in content than d_1, because d_2 includes d_1 as its history. Thus negation involved in the heterogeneous duration is not a pure indifferent 'is not' indicating a simple fact 'x is not y'. 'Is not', which underlies the identity of its elements, is immediately complicated as 'more than', 'later than', 'part of', etc., and these relations in themselves are involved in the formation of the heterogeneous duration.

Negation entails directedness of one element to another, comparison, thus playing a positive role –coordinating terms which it differentiates. In particular, past and future entities, albeit absent in the present, are not negated absolutely. They are not those which 'are not' but those which 'are not yet' or 'are not any more', as well as 'not here'. This in its turn indicates that real qualities have some directedness built into them, a vector value, so to speak, and that their being, even in relation to itself, includes being in a particular place and in a particular period of time, because outside that place and that time the entity in question does not exist in its embodied form.

Qualities are not locked in within themselves but reach out for other qualities, transcend themselves, emanating relations or, in other words, are in themselves vectorial: they harbour relations with other qualities within their own nature: if x is later than y, then x must bear within itself an in-built vector which directs x towards y and makes them comparable. In particular, when qualities are present, the 'presentness' is both a relation of simultaneity with other active phenomena and an intrinsic state of the event. It is down to the inherent structure of the process ontologically to assert that it is active and present, but being present is also a relation directed outwards of oneself, revealing the temporal vector of the given phenomena.

It may also be said that negation is the most basic relation that individuates qualities within duration when we fail to do so via ostensive affirmation. Whist we cannot establish the exact positive parameters of, say, note d within *abcda*, we can still say that d is not *abcda*, or that d is not a or c. We could summarize the above discussion by saying that the way in which elements can be distinguished in a non-numerical heterogeneous multiplicity is by using what we call here constructive negation that helps to expose the identity of qualities. It shapes qualities, by opposing their sameness to the surrounding otherness, but it also binds the same and the other in various relations.

Without introducing this kind of negation into ontology, we end up with every feature of reality being omnipresent, which would, in its turn, endanger the concept of heterogeneity, homogenizing all content, making no distinction between presence and absence, between the past, present and future. Eventually, we would have to accept everything as some kind of undifferentiated One, into which all differences would inevitably collapse. Allowing constructive negation as a relation that being generates, we are able to maintain the balance between presence and absence, and comprehend temporal succession: for example, despite the difficulty of finding positive characteristics for the present, it can be characterized negatively as not past and not future.

In heterogeneity, the same and the other are distinct qualities connected together, but the connection between them is not a loose relation but an in-built feature which assures the integrity of the entire heterogeneous whole. As Bergson warns, if we separate terms of this whole and allocate to them fixed temporal and spatial coordinates, we will erase the connection between them, which in its turn, would fragment the heterogeneous whole, reliant on the integrity and balance of all its terms. It could be said that in a heterogeneity, there are no gaps between qualities, and no mediating elements apart from qualities themselves. Their relations are made up of vectors found in qualities, and the joining is complete. However, the differentiation between qualities is maintained by constructive negation which opposes the same and the other within heterogeneity.

Thus we could argue against Bachelard that the idea of fullness does not contradict the qualitative diversity of being: there can be fullness without emptiness, but the sameness of one quality implies its absence in the otherness of another quality. Bachelard seems to imply that absence of a quality, even if replaced by the presence of another quality, means a gap, a lacuna. Does this have to be the case? The question is: what is the ontological status of non-existence and absence?

The Bergsonian duration, as we can see, progresses and expands with time. Its content is constantly on the increase, and new qualities are being added to the previous ones. In this process we do not see any gaps between the old and the new: the old simply grows bigger, and the new does not appear at the expense of the loss or absence of the old, because the old does not disappear in order to give way to the new. As time expands, it makes room for both. Where we do, nevertheless, see a gap is looking at the old prior to the new. In our next task of applying heterogeneity to selfhood, we will observe these gaps in trying to find a foundation for psychology in living matter, and a foundation for biology in inanimate matter. Granted that time is a constructive component of duration, it is not however the only constructive component. In order to give rise to a new level of evolution, time alone is not enough. However, new layers of being appear in the new time, and create a relation of their own absence in the past. It is these lacunae that we shall find difficult to fill or explain, in order to preserve the idea of continuity. Constructive negation explains the diversity within a unity, but it does not explain how new types of being appear seemingly from nowhere.

3   Changing the Past: Backward Causality

In our opinion, the philosophy of duration unavoidably leads to an observation that the future affects the past. Bergson comes very close to saying
this in his later philosophy: in chapter 3 of *The Creative Mind*, entitled
'The Possible and the Real', he argues that we can say that such and such
a real event was possible only in retrospection, for it is the realization of
it at some present time that made its being possible or potential in some
past time. For, as the future is essentially uncertain, the state of affairs in
the present does not entail the possibility of being a kind of pre-existence
of some future events which will eventually happen. The following quo-
tation is almost an introduction into the philosophy of changeable past:
'Backwards over the course of time a constant remodelling of the past by
the present, of the cause by the effect, is being carried out', because 'it is the
real which makes itself possible, and not the possible which becomes real'
(*CM*, p. 104). Bergson stops at this but we will argue that the past (not
as part of the metaphysical framework but as part of embodied temporal
reality) constantly changes,[17] affected by the events that follow.

We remember from *Matter and Memory* that for Bergson, the past
temporal reality is primarily disembodied reality. It is the reality that 'acts
no longer' (*MM*, p. 68), unknowable when taken as the pure past (*MM*,
p. 135). Yet the past survives (*MM*, p. 149). All of the past pushes onto the
present and some of it is recalled to life – in a new guise, as a foundation
and backing for new, present occurrences.

Our key claim regarding the past as part of heterogeneity will be that
the past changes, and the proof will be based on the role that retrospective
issues play in the construction of the real. Naturally, we must expect that
any talk of changing the past may provoke an extremely sceptical reaction.[18]
Indeed, changing the past is a taboo imposed on us by the laws of physics,

---

17   For an alternative treatment of possibility and retrospective issues see Mullarkey,
      Bergson and Philosophy, 68–81.
18   For a meticulous and thorough validation of the irreversibility of temporal pro-
      cesses see Adolf Grünbaum, 'Time, Irreversible Processes, and the Physical Status of

logic and ontology, and we find that all philosophies agree that the past is written in stone once and for all. When imagination dares to challenge this taboo, the attempts are severely punishable – Faust, meddling with the natural course of biological processes, must hand over his soul to the devil in return for his second youth.[19] Yet we intend to prove that the idea of duration requires our accepting that in some sense the past of temporal reality can and does change, and even that the foundation for this view exists in Bergson's original texts, albeit without his being aware of the fact.

What we mean by the idea of changing the past needs to be clarified. Ordinarily, this would be understood in the terms of science fiction as going back to the chosen moment in the past and altering the physical content of events that happened then. We mean nothing of the sort. On the contrary, time travel is an ontological absurdity, from the Bergsonian point of view, because leaving one's original temporal coordinates would be as destructive for one's identity as leaving behind one's size and one's colour.[20] Time, for Bergson, is the essential stuff everything is made of and, of course, it appears as particular temporal stretches, integrated into the network of global time, not as some loose chunks of time which hang by themselves, unattached to the total temporal continuity, and which can float from one set of temporal coordinates to another. Besides, if the travelling entity, whilst entering the past, retains all its experiences and memories acquired in consequent times, then we must assume a multiplication of timelines, where one present coexists with another present, which is really past: we take it for granted that these facts are unchangeable. Neither do we intend to claim that events can exchange their temporal positions, or be erased from the past. What we are proposing is that in temporal heterogeneity, where all elements are bound together, later elements contribute something to the content of earlier elements.

Becoming', in J. J. C. Smart, ed., *Problems of Space and Time* (New York: Macmillan, 1968), 397–416.

19 Johann Wolfgang Von Goethe, *Faust* (Oxford: Oxford University Press, 1998).

20 For an earnest consideration of the possibility of time travel see David Lewis, 'The Paradoxes of Time Travel', in Robin Le Poidevin and Murray MacBeath, eds, *The Philosophy of Time* (Oxford: Oxford University Press, 1993), 134–46.

Let us consider the following two events:

1 The assassination of the Archduke Franz Ferdinand of Austria and his wife on 28 June 1914 in Sarajevo by Gavrilo Princip.
2 The First World War.

Historians universally agree that event 1, the assassination, is a cause of event 2, the war, and therefore we will treat them as links in the objective chain of events. When 28 June 1914 was the present time, and the Archduke had just been killed, it was a formidable but localized event. No one, including the assassin himself, would have seen this as the start of a war which by August would already involve Germany, Austria-Hungary, France, Russia, Serbia, Britain and Belgium. In June, the recent assassination was a mere act of personal terrorism, a political murder. But in August it became the cause of the First World War, which later involved other nations as well and cost the world an estimated ten million lives.[21]

Thus event 1, the assassination, became something else – the cause of the First World War – after it was no longer present. We cannot say that the assassination was the cause of the war at the time of it taking place but we just did not know about it, because on 28 June 1914 the future state of affairs was uncertain and the assassination may not have led to the war if, for example, Russia had not mobilized in response to the Austro-Hungarian ultimatum to Serbia. But the events that followed the assassination did lead to the war, and so events which followed event 1 changed the nature of the event 1, by contributing new content to it.

To take a biographical, example, Wellington's biographer, Richard Holmes, in his account of the Iron Duke's earlier years, mentions: '... [H]e was soundly thrashed by a young blacksmith named Hughes, who was proud to relate how he had beaten the man who beat Napoleon.'[22]

Holmes brings this up as nothing more than an amusing anecdote, but we see in this the work of time that makes changes in the past. The youngster whom Hughes had thrashed was not, of course, the Duke of Wellington

---

21   Susanne Everett, *World War I* (London: Hamlyn, 1980).
22   Richard Holmes, *Wellington The Iron Duke* (London: HarperCollins, 2002), 8.

who celebrated the victory at Waterloo. He was not even the future Duke of Wellington, because when the scuffle with the blacksmith took place somewhere in the 1780s, the future did not shape itself favourably for Arthur Wellesley. Far from expecting him to become a significant figure in the history of Europe, his family hoped that he would merely get by in life, so to speak, and a free commission in the army was not the preferred but the only option available to the sixteen year old of no special talents or fortune. Such were the present situation and the future projection of it for Arthur in the mid eighties.

Thirty years on, when the present time falls on the year 1815, Arthur Wellesley is Duke of Wellington and a national hero. We can see how the embodied eventuality of 1815 changes the nature of an event which happened back in the 1780s. By 1815, the Arthur Wellesley of the 1780s in 1815 had become the future Duke of Wellington who would beat Napoleon. So it is only retrospectively that Hughes the blacksmith became the one who had beaten the man who was to beat Napoleon, and not when the scuffle between the young men actually took place.[23]

In *Creative Evolution*, Bergson refers to the human species as 'the culminating point of the evolution of the vertebrates' (*CE*, p. 141), on the basis that it claims 'the entire earth for its domain' (*CE*, p. 140). If we agree that the dominant species ought to be considered an end of evolutionary development then humans are indeed its culminating point. On the other hand, this is the state of affairs at present. We can imagine a remote future where humans will have evolved as a new species, creatures, say, with a far advanced brain capacity who will claim not only the earth but other planets as well. In that remote future the humans of the present day will become past, and no longer will be the end point of evolution. They will have become a stepping-stone towards a more developed, more powerful species, and so their role in the universe will have changed after they have gone.

---

23   For an alternative explanation of this, let us say that the researcher who is writing these words started school in 1982. In 1982 that child was not the future researcher because the future is generally uncertain and she could have chosen another career. But now she is a researcher, and this fact makes the child of 1982 a future researcher retrospectively.

Thus we shall insist that the past changes and it is the causal relations that are best suited to reveal its changeability. This can be presented schematically in the following terms. Let us assume that c is a cause and e is its effect, and that c is earlier than e. When c takes place at time t, it is not yet the cause of e, because e has not yet happened, and we accept that the future is ontologically uncertain. c becomes the cause of e when e takes place at time t', but by that time c is already past. Thus the characteristic of c as the cause of e is attributed to c in retrospect – c becomes the cause of e when c is already gone. Our claim here is that e has changed the past by attributing to c a new ontological feature – that of being the cause of e.

We shall try to anticipate possible arguments against this. For example, we may be questioned on the very concept of causality. For example, how do we distinguish between a true cause and a mere condition of something happening? We could quote Collingwood who states that both would be a part of the genuine cause, which is 'made up of two elements', and '[n]either ... could be a cause if the other were absent'.[24] But we shall go further than that. Remaining on the Bergsonian premises, we can cancel out the distinction between cause and condition because, for Bergson, the emergence of the present is caused by the entire past, and as both cause and condition are past, it makes no sense to distinguish between them. But going even further, from the platform of the revised Bergsonism, we shall allow the distinction between the decisive cause of an event and the mere condition which made it possible. We shall say that the difference between them is temporal. The state of affairs at time t is a condition for a range of future possibilities at the future time t', but when the future time t' becomes present and the reality has shaped itself as a definite present event, then this condition has become the cause and as such, gained a new causal property.

Also, a sceptic would say, it is not the past that changes, but the present contains a new evaluation of the past which makes no difference to the past itself. He or she may add that causality is a relation which matters to the effect and not to the cause. In response to that we can say that it is quite right to characterize causality as a relation: we have already introduced

24   R. G. Collingwood, *An Essay on Metaphysics* (Oxford: Clarendon Press, 1969), 292.

it as a relation ourselves. However, our analysis of qualities and relations within heterogeneity[25] convinces us that they do not come in a pure form, autonomous and independent of each other. Qualities inevitably prolong themselves as relations, and relations in their turn affect qualitative properties to varying degrees. This is true because qualities and relations do not hang suspended in mid-air waiting to be connected but rather produce each other, keeping up the idea of continuity and duration. So when we talk of causal relations, we find that for e it is qualitatively important that it is an effect of c because this fact reflects its history, but then it is equally qualitatively important for c that it is a cause of e, for c cannot remain indifferent to a relation that connects it with another quality: relations connect both terms, and are not one-sided. But that new qualitative addition is only possible after the emergence of e, i.e. when c is past.

Another criticism could be that it is a logical condition of c that it is the kind of quality that can give rise to e, and that this characterization eliminates the point which we are arguing. To this we would reply that firstly, Bergson already says that the potentiality is determined retrospectively, after it has been realized. Secondly, and more importantly, there is an ontological difference between *can* and *did*, and the quality that *can* give rise to e is not the same as the quality that *did* give rise to e. It may be argued that if the assassination of the Archduke had not happened, the war would still have broken out, triggered by something else and therefore the assassination is not the actual and ultimate cause of it. But due to the ontological difference between *could* and *did*, the fact is that Gavrilo Princip's actions led to a chain of events which escalated into a global conflict, however bizarre this may seem when we compare both events. Alternatively the sportsman who did win the trophy is rightly treated differently from all those who could have won the trophy but never took the trouble to fight for it. Thus time, we insist, affects not just the embodied reality which it constructs but also the embodied reality which it leaves behind.

It may also be argued that if we say that e is a cause when c happens as its result, this only changes our subjective interpretation of past events

---

25   See Section 1 of this chapter.

which does not necessarily reflect the objective state of affairs. An answer to that could be the Bergsonian reply that interpretation is also part of the being of the event. But to make it stronger, we will say that if the division of reality into causes and effects is subjective and ontologically unreliable, then any continuity falls apart into atomized elements and dissipates altogether because, if causes and effects are dispensed with, there is nothing to hold the continuity together objectively. Every minute event can be divided into causes and effects, and if we regard causality as superimposition of interpretation, then even minute events will be eliminated from ontology. With every phenomenon broken down to unconnected elements, we would end up with a picture of an atomized, pulverized universe, but even that view could be said to be a result of a subjective interpretation.[26]

As for Bergson's own inadvertent contribution to the theory of a changeable past, we get glimpses of the constructive role of retrospectivity as early as Chapter 2 of *Time and Free Will*, where Bergson claims that duration 'forms both the past and the present states into an organic whole' (*TFW*, p. 100). Bergson unpacks this later as a unidirectional process, with the movement being effected solely from the past towards the future. However, we are compelled to say, if past and present states melt 'into one another', then the influence should be mutual, in the way that not just the past elements would affect the present ones, but the present ones should somehow affect the past ones.

Next we learn that it is retrospection that underpins and constructs spatial representations. When we perceive a succession of elements we, according to Bergson, set them out in line '*after* having distinguished them' (*TFW*, p. 102, italic mine). For Bergson, setting out in line is spatializing, which is a wrong and misleading procedure, but we see in this the positive work of arranging past elements into something new that can only appear retrospectively.

---

26   For an alternative defence of the objectivity of causal relations see Michael Dummett, 'Bringing about the Past', in Robin Le Poidevin and Murray MacBeath, eds, *The Philosophy of Time*, 117–33.

In Chapter 3 of *Time and Free Will* Bergson portrays retrospective self-consciousness which reveals to our present self the motives of our past actions of which we were not aware at the time: '[W]e shall believe that we are acting freely, and it is only *by looking back to the past*, later on, that we shall see how much we were mistaken' (*TFW*, p. 169, italics mine). The same would apply to the understanding of one's own free acts: we realize that we acted freely, only when turning back to our past ideas and feelings, 'not unperceived but rather unnoticed at the time of acting' (*TFW*, p. 169).

Further on Bergson scrutinizes the schema of choice between paths OX and OY (*TFW*, p. 176): '[I]f I dig deeper underneath these two opposite solutions, I discover a common postulate: both take up their position *after* the action X has been performed' (*TFW*, p. 179, italic mine). In what Bergson negatively calls spatialization, I see the constructive work of retrospection (see Diagram 2, p. 22).

In Chapter 2 of *Matter and Memory*, there are more unintended references to constructive retrospectivity. When Bergson asserts that 'complete perception is only defined and distinguished by its coalescence with a memory-image' (*MM*, p. 127), he also mentions that 'the memory-image itself, if it remained pure memory, would be ineffectual' (*MM*, p. 127). This is because '[v]irtual, this memory can only become actual by means of the perception which attracts it' (*MM*, p. 127). What matters to us here is Bergson's acknowledgment that a later event (perception) affects the previous event (memory) by bringing it to life. This said, we understand that even after being brought to life, the memory remains what it has been – a memory, a past event.[27]

---

27 Memory is understood here in the Bergsonian sense as the accumulated past.

4   Predetermined but Changeable Future

Despite Cunningham insisting that considerations of the future are abhor-
rent to Bergson,[28] the idea of the future can be retrieved from his arguments
against finalism. Bergson argues that, in the existing tradition of thought,
objective things, from Plato's Ideas to Kant's noumena, are considered outside time. This implies that everything already exists, but the limitations
of our cognitive ability prevent us from seeing everything as already existing, making us perceive everything as becoming, unrolling *in time*. Thus
temporality is superimposed on the real by our subjectivity and is seen as
superfluous to the real being.

Bergson asks: if everything is already given at once and the future is
thus predetermined, why does it take a certain amount of time for events
to unroll? His own answer is that time is not a mathematical construct that
can be reduced, stretched or eliminated completely from the real events.
Time is real, absolute and irreducible as it *is*, and it plays a constructive
role in the process of becoming. Time is 'causally efficacious' (*CE*, p. 41)
for the following reasons.

1   The time it takes for a process to unroll is irreducible and is an integral
    part of the makeup of that process – just like its physical components.
    It takes a certain length of time for a process to happen; this *time*
    cannot be eliminated from the process, or reduced: therefore it is not
    relative but absolute.
2   Each moment of time succeeding the previous one has never occurred
    before and will never occur again, and is in this sense absolutely new
    and unique. It always takes a *new, different* period of time for a new
    process to happen: thus time ensures novelty.
3   Time is not an empty container being gradually filled with the eventual
    content. Events 'make' time as they unroll, and constitute time itself.
    Thus the newness of each portion of time automatically becomes a

28   Cunningham, *A Study in the Philosophy of Bergson*, 130.

property of its content. So the content can be apparently the same as in the past, but it is nevertheless different because it happens at a different time.

4 In duration, loaded with the past in the form of memories and experiences, any repeated experience affects each time a new, older duration, resulting each time in a totally new state of the being in question because it is a different being.

Thus the Bergsonian future can be described as predetermined in form (in that everything will grow old and that all future events will be new) but indeterminate in content. However, we fail to see how the efficacy of time eliminates the possibility of the future being predetermined in the Platonic sense, if we take determination and novelty in the Bergsonian sense. There is a sense in which determinism denies novelty if novelty is to be understood as reality with original content, never thought of or encountered before. But for Bergson, the very fact of something existing in a new present entails originality and novelty and, from this perspective, his own claim that a plan would forfeit novelty carries no weight, because the realization of a plan would unroll in a different time than the creation of the plan. The plan is an idea; its realization is a series of events with their physicality and temporality, predetermined by the plan but not equal to the plan. The fact that the plan predetermined the events makes the former a cause and the latter an effect, but does not equate them.

There is also the hard-line mechanistic view of predetermination as mechanistic causation,[29] which Bergson plainly reduces to predetermination as a plan and hence does not address it separately. 'The essence of mechanical explanation ... is to regard the future and the past as calculable functions of the present, and thus to claim that all is given' (*CE*, pp. 39–40).

---

29 For instance, according to Skinner's position, if we abandon myths about freedom and choose the scientific view of reality, then we would inevitably admit that we have no capacity to interfere with casual properties of matter (B. F. Skinner, *Science and Human Behaviour* (New York: Macmillan, 1963)).

The revised Bergsonism from our standpoint allows us to account for both predetermination and novelty, and also challenges the mechanistic view of rigid determination. We suggest a partially predetermined future, which changes with every new present. The future possibilities cannot be totally indeterminate – they must be compatible with the previous development – and, however numerous and varied, are limited, and as such they are ontologically predetermined to a certain degree. It is the past – not the plan – that limits future possibilities. One does not need a plan to predetermine the future – the development of things themselves does it – but, although this renders the universal plan unnecessary, it does not dismiss it either.

If there is a plan, it may be produced *internally* as a logical projection of the actual state plus a certain tendency, coming from the past, which predetermines what will happen next, and after that, etc. Then there would be a future, predetermined by firstly, the past (inasmuch as it contains the decisive tendencies) and the present, and secondly, by the fact that it is inevitable that something will *be* in the future. So, if in a current state of things we can see what will become of it next, there is a plan as an anticipation of the future. But the realization of the future in the present will affect the state of affairs, and then the plan for the more remote future will change, as it will be a projection of different circumstances: so, *the predetermination itself will change*. It is only in the present that events and things take their final and definite physical form and become what they are. It is the present that is determined in such a way that it cannot change, but the future, a projection of the present, is vulnerable to change and infinitely flexible – until it becomes present. Our key claim here is that *different presents have different futures*.

In Bergson, the reality of the present comprises all of its history, so we can say that the next stage following every present state is predetermined by all of its past, all of the content of its being up to the present moment. So, a thing as a process completed up to a particular moment has its future determined and limited by all of its content. The thing at time t, comprising a content of *abc* has its future lying ahead of itself, determined as *abcdef*, destined to have occurred by time t', and *abcdefghi*, destined to have occurred by time t'', as in Diagram 5.

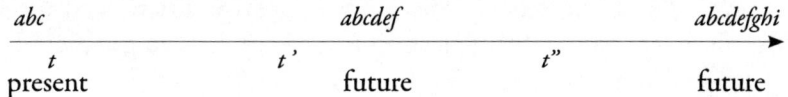

Diagram 5 The state of affairs at time t

Diagram 5 illustrates the state of affairs at time t, with the state *abc* being the history of an object up to this moment. The state *abcdef* is a development of *abc*, where the *abc* element will be retained and *def*, added on by time t'. As a further development of *abc*, the state *abcdefghi* will have occurred by time t''. So, at time t, the future states *abcdef* at t' and *abcdefghi* at t'' are determined by *abc* at time t'.

However, the state *abcdef* at t' is a new reality, and at t' it is *abcdef* that will determine the future state at t'', and not *abc*. A new, different reality with new content, *abcdef*, will determine a different future from the old reality, *abc*. So, at time t' the future state at t'' is altered compared to what it is at t. At present time t' the future t'' may be, for example, *abcdefhik*, as illustrated by Diagram 6.

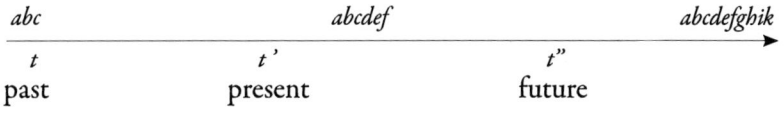

Diagram 6 The state of affairs at time t'

The hard-line mechanistic view would still hold, however, that this state of affairs is already predetermined at time t. But according to our theory of backward causality, *abc*, which at time t implicates *abcdef* and *abcdefhik* at time t' will have a different identity, enriched by relations with *abcdef*. Therefore its prolongations into the future existing at time t will become altered at time t'', and this, we believe, effectively disproves the mechanistic view of reality where all future possibilities are already entailed in the past.

So, the future exists as predetermined by the whole of the accumulated temporal content, and is altered every time some new reality is added on to the whole of the existence of things. The future is predetermined by the past, but the content of this predetermination keeps changing because the past keeps changing.

5   The Present: Temporal Shift Within the Self

The embodied present as part of the heterogeneous whole involves the following controversy. It is the criterion of embodied reality: real events must at some stage be present. But when we attempt to focus on the present moment, the present time slips through our fingers, and avoids the grasp of our thought.

The embodied present is a quality within a heterogeneity which we cannot demonstrate ostensibly but which manifests itself via relations with other members of the heterogeneous whole. The reality, including our own existence, hinges on the present: both the future and the past are refracted through the present and are changed by it. We cannot define the boundaries of the present from the standpoint of the present itself, but we say, using constructive negation, that the present is not past and not future. This eliminates what the present is not. Also, the present is not just 'not past', and 'not future' – it is not past *yet*, and not future *any more*. The present entails vectorial properties which locate it within temporal continuity: its position between the past and the future is not a superimposed relation but is entailed in its own qualitative makeup.

We cannot rationalize our experience of the present in that very present, because the present passes too quickly. Instead, we seek evidence of it in the time which is either past or future in relation to it. We know that the future will come as present, and anticipate its content more or less accurately, because the existing reality projects itself into the future as a limited range of possible outcomes. On seeing a falling object, we anticipate its hitting the floor in the coming future moment. But no sooner have we focused our attention on this situation, than we have already lived through the present time of the fall, and now it is the past time of the fall. However, the outcomes of that formerly present event – the object lying on the floor plus our memory of the recent fall – indicate that the embodied event of the object falling has just been present.

We observe a certain temporal imbalance, a shift of temporal planes, between our conscious act and the objective process towards which this act is directed. Our physical and sensory existence, together with other material objects and organisms, joins in the formation of the present. Our thought, however, cannot access that present while it is present. Our conscious attention is either behind it, or in front of it, depending on whether we anticipate it or reflect on it: our awareness misses the present time of the physical reality of the world.

The crucial moment here is that our awareness also misses the present of the physical reality of our own bodies inasmuch as they are part of the physical reality of the world. What Bergson labelled a spatialization of the fleeting reality by intellect may be due to the temporal displacement between the consciousness and the body within a person. The rationalizing mind needs a distance from its objects to perform its act, and is forced to maintain a temporal distance between itself and the embodied reality, including its own body. When the body lives in its present, the consciousness accompanies this process from the premises of its own present which is either past or future in relation to the present of the body. In an attempt to double on itself and reflect on its own present, the mind has to abandon this present as well and access it from outside, temporally removed from it. Thus even the present of the mind itself is not accessible by the mind whilst this present is present, because the temporal act of rationalisation requires its own time to unroll, and this time does not coincide with the time of its object.

The complete coincidence of thought and its object, which should take place in the Bergsonian intuition, is also impossible if we revive Bergson's implicit observation that perception and, we shall add, acknowledgment of any type, is a process and as such, is stretched temporally. If it is triggered by the object at time t, it takes a stretch of time for it to develop into an image shared by the object and the perceiving subject, and for the subject to have thoughts about it. Thus the image will always be temporally dislodged and stretched in relation to the original object, as it has to belong to the time of both the object and the subject. What we attempt to think of as present is, in fact, past, or the present image of past reality. As this past reality includes our own bodies, the images that they emanate are delayed in relation to their actual existence.

When we direct our mind to an unrolling event, we see a continuous, unbroken development and expect that if we focus hard enough, we should be able to connect with the very present time of the observed process. Instead, our conscious attention fluctuates between the reality that has just been present and has just shaped itself, and the projection of this reality into the nearest future. Our consciousness misses the phase of the actual shaping of reality and catches up with it only when it has already been shaped, which means that the present of embodied reality is always either anticipated by our mind or delivered to our mind with a slight delay, and both the anticipation and the delivery happen in the present time of the mind which does not coincide with the present time of the world and the body. Thus mind and body do not exist simultaneously but are in a fuzzy temporal field where consciousness in the present corresponds to the body in the past and in the future.

Suppose our self is at time t'. Both the mind and the body and the rest of the world are present at the same time t'. Our mind is able to grasp in experience or in thought the body and the world at the previous time t and at the following time t'', but is unable to connect with the actual time t'. The illusion is that the mind is grasping the present in its experience, but what we see is a flyover of consciousness from the future to the past and from the past to the future, and the state of unconsciousness corresponding to the present.

To support this view of the present, we shall recall the Augustinian confession of the impossibility to rationalize time: 'I know well enough what it is, provided nobody asks me; but if I am asked what it is and try to explain, I am baffled.'[30] Geoffrey Bennington interprets this as the inability to *focus* on time and links this problem to Lyotard's discussion of lateral vision whereby the attempt to bring lateral vision of an object of investigation into focus loses the object by transforming it into the focal vision which it is not meant to be.[31] Bennington suggests that time may be simply lost in a thematic presentation and proposes philosophical acceptance of 'the unknowledgeable knowledge suggested by Augustine'.[32] We take a similar attitude in respect to the present time, the time that passes right now, and suggest that it is never in focus but always in a lateral field of our mental vision. When we manage to bring it into focus, it is no longer present but is shifted towards the immediate past or the immediate future. Also, George Rostrevor, in his analysis of the Bergsonian intuition as the cognitive grasp of one's being coinciding temporally with one's being, finds the same inner temporal split: '[W]hen I examine this intuition, I am led irresistibly to the conclusion that the reflective element is not only present in it, but is in truth the only active element present. When I am completely absorbed in some activity, I have not *at the same moment* any consciousness of the nature of that activity'.[33]

---

30 St Augustine, *Confessions*, 264.

31 Geoffrey Bennington, *Lyotard: Writing the Event* (Manchester: Manchester University Press, 1988), 73–4.

32 Geoffrey Bennington, 'Time after Time', *Journal of the British Society for Phenomenology*, Vol. 32, No. 3 (October 2001), 301.

33 George Rostrevor, *Bergson and the Future of Philosophy: An Essay on the Scope of Intelligence* (London: Macmillan, 1921), 55.

Time

1   Time as a Quality and as a Relation

In this chapter we shall consider specifically time and its role in duration. Bergson's major achievement in the understanding of the nature of time, as we see it, consists of maintaining and proving that time belongs to temporal reality as its integral part rather than something that exists separately from phenomena.[1] He successfully internalized time into the reality of things but in order to fortify his position, Bergson opposed time as part of things to time as a framework for everything that exists, dismissing the latter as a false, spatialized view on time.[2]

Whilst declaring time to be a qualitative, essential element of things, Bergson makes it impossible to consider temporal relations between things, because it can be said that they are nothing but spatial features where elements, for the sake of comparison, have been taken out of the genuine temporal sequence and presented simultaneously. As it is clear from Bergson's equating the reality of time with its efficacy,[3] from his refutation of the possibility of measuring time since measuring would involve comparison of

---

1   Kolakowski sums up Bergson's philosophy in one sentence: 'Time is real' (Kolakowski, *Bergson*, 2). An opposite claim that time is unreal is represented by McTaggart: see John McTaggart Ellis McTaggart, *The Nature of Existence*, Vol. 2 (Cambridge: Cambridge University Press, 1968), Chapter 33 'Time', 9–31. For refutations of McTaggart see D. F. Pears, 'Time, Truth and Inference', in Anthony Flew, ed., *Essays in Conceptual Analysis* (London: Macmillan, 1966), Chapter 11, 228–4, and Andros Loizou, *The Reality of Time* (Aldershot, UK: Gower, 1986).

2   For an exposition of this, see Chapter 4, Section 1.

3   See Chapter 5, Section 9.

temporal stretches and their simultaneous alignment,[4] and from his claim that duration can only be comprehended from within,[5] we can understand that any relation between temporal stretches would be dismissed by Bergson inasmuch as they imply an externally initiated simultaneous alignment of timeless entities. Thus, according to Bergson, relations between events such as 'earlier' and 'later' would merely be a spatialized view of time, where events are wrongly taken out of the genuine temporal sequence, mentally juxtaposed as if they were simultaneous, and their positions compared, very much like the spatial positions 'nearer' and 'further'.

I understand that, for Bergson, to be real is to be a quality, not a relation; that a quality must be a property that intrinsically belongs to an entity, whereas a relation is a property superimposed onto an entity in an epistemological act of a conscious mind, which mentally aligns one entity with another. A quality is a constructive component of a phenomenon, and exists in its own right, whereas the existence of relations is secondary to the existence of self-sufficient qualities. A relation emerges as a resultant of a combination of at least two terms; it is a derivative of this combination and therefore its ontological status is less firm than that of a quality. Dependent on other things, a relation seems to be less stable, less dispensable and therefore less real. To defend the reality of time, Bergson insists that time is a quality and not a relation, thereby confining real time to the inner reality of things.

Bergson's position on this is not unlike Bradley's.[6] Alongside external relations, which are superimposed on terms and do not affect their inner nature, Bradley introduces into philosophy internal relations, which connect the properties of an entity and reflect its very make-up. Eventually, he considers all relations to be internal, because he finds that the nature of things and of the whole universe is defined by this global network of relations. For Bradley things are not just characterized by relations – relations capture the inner nature of things. Every relation 'exists within, and

---

4   See Chapter 2, Section 2.
5   See Chapter 2, Section 4.
6   F. H. Bradley, *Appearance and Reality: A Metaphysical Essay* (Oxford: Clarendon Press, 1978), 24–5.

by virtue of an embracing unity, and apart from that totality both itself and its terms would be nothing. And the relation also must penetrate the inner being of its terms'.[7] The assertion that all relations are ultimately internal is close to the Bergsonian view because Bradley is only one step away from saying that all relations are ultimately qualities. We, on the other hand, agree with G. E. Moore who counteracts Bradley's eagerness to internalize all relations, arguing that it is not true that all relations affect their terms.[8]

Taking Bergsonism further, we need to align Bergson's idea of temporal continuity, where interrelated temporal elements are in a state of mutual proximity, immediacy and interpenetration, and those elements which together do not form temporal continuity but which are nevertheless temporally related to each other as events that happened earlier or later than each other. Retaining the Bergsonian idea of time as a quality, we must also offer an explanation of temporal relations.

In this discourse, we accept internal relations as relations between qualities within heterogeneity, and external relations as those which exist between entities or events that do not affect each other's nature. We have already dealt with internal relations in Chapter 5, but it is more difficult to incorporate external relations into Bergsonism. If we do not, however, then Bergsonism remains a theory that absurdly refutes the ontological reality of 'earlier' and 'later' when it comes to components taken from distinct continuities, and maintains the impossibility of saying, for example, that the discovery of America took place later than the extinction of the mammoth.

The reason for this refutation lies in Bergson's implicit claim that a relation necessarily means a spatial relation, where genuinely temporal successive elements are falsely presented as simultaneous, because their bringing together in a relation involves their alignment in one plane, and thus they are made to coincide temporally, albeit in one's mind. The strict Bergsonian view thus implies that events which are separated in space and time are not related at all, and constitute a mini-universe each – unless, of course, we accept the universe as a giant organism.

7   Bradley, *Appearance and Reality: A Metaphysical Essay*, 201.
8   G. E. Moore, *Philosophical Studies* (London: Routledge & Kegan Paul, 1965), 277.

Contrary to Bergson, relations do not necessarily presuppose a simultaneous alignment of qualities, whereby qualities are given, and relations are imposed on them, which, for Bergson, amounts to spatialization. There are genuinely temporal relations as well and the temporal relations of 'earlier' and 'later' are not reducible to spatial relations. The juxtaposing and spatializing of 'earlier' and 'later' would be merely a result of a particular usage of language, such as in tenseless theories of time.[9] Tenseless theorists argue that it is possible to disregard tensed statements and account for temporal relations by saying 'is earlier than' and 'is later than'. Instead of saying 'x is past' and 'y is present', they suggest 'x is earlier than y' thus avoiding the supposedly subjective view of reality and asserting ontologically eternal 'timeless' truths. The element 'is' indicates that temporal relations are fixed in eternity as, if x is earlier than y, it is always earlier than y. Bergson would argue against such a way of seeing relations because the 'is' used in the present tense aligns temporally diverse x and y in one time – the present. This way of talking also implies that reality and all relations in it are fixed from the start, as the tenseless 'is' expresses necessity. However, instead of dismissing temporal relations, we can say that since reality is not fixed, x does not become earlier than y until both x and y become real, so at different times x will be, is, and was earlier than y, because, for example, when both x and y are in the past, they 'were', and their relations with each other also 'were'. A similar argument is used by C. D. Broad against the timeless usage of 'is' in McTaggart,[10] but we argue here against its present use.

Even though in our mind we may juxtapose events in a quasi-spatial simultaneity in order to ascribe to them characteristics of 'earlier' and 'later', in reality events undoubtedly *happen* earlier and later than other events. 'Earlier' and 'later' are not relations which are either fixed a-temporally or belong to the present in which events are compared; they *emerge* as events unroll and are irreducibly temporal relations. This shows that, firstly, the related items are not a-temporal spatial nodes but are events in themselves,

---

9   See McTaggart, *The Nature of Existence*, Vol. 2, 9–31.
10   R. M. Gale, ed., *The Philosophy of Time* (London and Melbourne: Macmillan, 1968), 136.

and, secondly, that the relation is not spatial, or timeless, but grows out of the development of things.

Accepting the Bergsonian argument that there are no timeless qualities in reality,[11] we rule out the possibility that temporal qualities may be reducible to temporal relations.[12] But we also need to show that the temporal relations of earlier and later are not reducible to temporal qualities. The key Bergsonian argument in favour of the claim that time is a quality is his proof that processes, taking time to unroll, need exactly that much time, not more or less, for their authenticity. This time – the particular length of it – cannot be dispensed with under any circumstances, if the identity of those processes is to be preserved. If this is true in respect to time, internal to an event, does the same apply to temporal relations between events?

Supposedly, using the Bergsonian example, it is essential to the solution of sugar that the lump of sugar takes, say, two minutes to dissolve in water.[13] These two minutes are irreducible and Bergson is convincing here. But does it matter for this process that it took place, say, twenty minutes after the news programme and ten minutes earlier than the film? Or would it be different if the news was broadcast after the sugar experiment, and the film was shown before it?

If we take the universe as a whole, then every minute detail matters because it defines its precise identity – and this correlates with the Bergsonian view. But this also means that the identity of every single event in the universe must be defined by all other events, and we will argue against

11   As explained in Chapter 3, Section 1.
12   For various discussions of temporal relations see, for example, Graeme Forbes, 'Time, Event, and Modality', in Robin Le Poidevin and Murray MacBeath, eds, *The Philosophy of Time*, 84–95; D. H. Mellor, *Real Time II* (London: Routledge, 1998); Richard Taylor, 'Spatial and Temporal Analogies and the Concept of Identity', in J. J. C. Smart, ed., *Problems of Space and Time*, 96.

For discussions of time as a quality see, for example, Anthony Quinton's view on times and spaces as individuals (Anthony Quinton, 'Spaces and Times', in Robin Le Poidevin and Murray MacBeath, eds, *The Philosophy of Time*, 203–20). On taking the Bergsonian idea of time as process further, see Alfred North Whitehead, 'Time', *The Concept of Nature* (Cambridge: Cambridge University Press, 1982), Chapter 3, 49–73.
13   For the exposition of the sugar example see Chapter 4, Section 1, 51.

such an extreme understanding of heterogeneity. Many events and their timing are irrelevant to the identity of many other events. So temporal relations with events that are irrelevant to a given temporal process cannot be equated with the qualities of that process. Therefore, temporal relations cannot be reduced to qualities.[14]

2   Time and Space

In Chapter 2 of *Time and Free Will* we find an opposition between 'real duration, the heterogeneous moments of which permeate one another', time, and 'the external world which is contemporaneous with it', space, the two sides of the real which are aligned and give rise 'to a symbolical representation of duration, derived from space' (*TFW*, p. 110). Time and space are not reconciled even in motion: Bergson affirms that although a moving body occupies space, motion itself, the progress by which the body passes from one position to the other, 'occupies duration' and 'eludes space' (*TFW*, p. 111).

Bergson separates time and space to such an extent that they become incompatible and the ontological relation between them dangerously weakens. 'There is neither duration nor succession in space', Bergson asserts, as 'each of the so-called successive states of the external world exists alone' (*TFW*, p. 120). The danger is that if the reality is thus broken up into the temporal and the spatial which do not participate in each other, forming a unity, then the apparent uniformity and coherence of the whole of the Universe becomes fiction.

14   Interestingly, Bradley discusses temporal (and spatial) issues in similar terms and concludes that since time (as well as space) is both a relation and not merely a relation, it is not, strictly speaking, real and is merely an appearance. Our observation that qualities and relations do not exclude each other permits a more positive conclusion: that time is real and that there are temporal qualities and temporal relations. (Bradley, *Appearance and Reality: A Metaphysical Essay*, 30–6).

In *Time and Free Will* Bergson divides reality into incompatible domains, space and time, and in *Matter and Memory*, he wants to reconcile them, but he does that by declaring everything, including space and substance, to be ultimately time and movement. The reason why we can apply the same terms to consciousness and to the material world is that, he claims, space can be removed from our analysis of the material world with no damage done to the adequacy of our understanding of physical objects.

> In regard to concrete extension, continuous, diversified and at the same time organized, we do not see why it should be bound up with the amorphous and inert space which subtends it – a space which we divide indefinitely, out of which we carve figures arbitrarily, and in which movement itself, as we have said elsewhere, can only appear as a multiplicity of instantaneous positions, since nothing there can ensure the coherence of the past with the present. It might, then, be possible, in a certain measure, to transcend space without stepping out from extensity; and here we should really have a return to the immediate, since we do indeed perceive extensity, whereas space is merely conceived – being a kind of mental diagram. (*MM*, p. 187)

The dispensability of space from spatial objects seems to be Bergson's decisive step towards bringing together spirit and matter. He describes physical reality in the same terms as he described duration in *Time and Free Will*, considering, he claims, things as they are, and not as they are interpreted by our action orientated mind.

Just as with the refutation of relations, Bergson's dismissal of space is a downside of his otherwise largely positive work of defending the reality of time as an independent essence. Bergson believes he has overcome successfully the prejudices that cloud our understanding of time. The main prejudice, according to him, is that time is mistakenly confused with space. He sees his own achievement in managing to expose this confusion and, separating the spatial from the temporal, captures the essence of time in the notion of psychological duration.[15]

Bergson considers space and time in openly axiological terms. Everything that has to do with space is false and lifeless, and everything that has to do with time is genuine and ontologically valid. But time can

---

15 For the exposition of Bergson's refutation of space see Chapter 2, Section 3.

be comprehended only with the catalytic involvement of space, as without spatial markers there is no way of distinguishing between earlier, later and simultaneous; between past, present and future; and even the direction of the temporal flow cannot be established without spatial orienteering. Contrary to some of the Bergsonian claims, the elimination of space from ontology and epistemology does not enhance the picture of reality, making it vibrant and constraint-free. We shall demonstrate that the downgrading of space that occurs in Bergsonism dangerously affects temporality and reality itself, as the reduction of spatial precision makes reality less temporal and therefore less real, inasmuch as temporality is a feature of real things.

Artists, whose task was to prove the temporal reality of the Biblical events, did so via symbolic depiction of time by spatial means, but the way they treated space affected the way they presented the time of those events and ultimately, their reality. Depicting the Baptism of Christ, Piero della Francesca (c. 1415–1492) wanted to convince the congregation that it did take place for real and in real time. (*The Baptism of Christ*, National Gallery, London). Working on this task, the artist obviously exploits spatial precision in order to achieve the sense of temporal reality. He pays meticulous attention to detail, as if indicating that when Christ was baptized, trees had the same leaves as those that one can see today, and so the Baptism must have happened in a real place and in concrete, real time. Also, Piero della Francesca deliberately transfers the scene of the Baptism from Palestine to Sansepolcro, his home town.[16] The point here is that the event took place in a town which looks like one's own, in a unique and precise place – a real place – therefore it must have happened in a precise period of real time. The artist provides an almost physical, photographical evidence of the holy event, and validates its historicity by using spatial precision, a necessary attribute of the present time and of reality which at some point must be present.

If the Italian painter's mission was to create an ocular demonstration of Christ's historical reality, icon-painters, by contrast, emphasized

16   Robert Cumming, *Annotated Art* (London, New York, Stuttgart: Dorling Kindersley, 1995), 19.

his divinity at the expense of spatio-temporal concreteness. For example, *Icon of the Descent from the Cross* (Late fifteenth century, Northern School, Tretiakov Gallery, Moscow) shows anatomically inaccurate human figures, simplified background, and a limited choice of colours. It may seem that the artist has failed to demonstrate that the depicted event ever happened, because if it had ever been embodied and present at some point, the scene would have looked more realistic.

But this does not matter, since the aim here is not to prove the historical veritability of the holy event, but to capture its holiness. The question whether the Descent did or did not take place in the chain of historical events is irrelevant: what the icon shows is that the Descent is not an ordinary, earthly event which took place in some concrete present and which is now past, but that this is a divine, spiritual happening, detached from any mundane empirical concreteness and possibly even from any particular here-and-now. It is assumed to have happened in actual fact, but once it had happened, it gained a property of a universal significance and omnipresence, being forever relevant, forever remembered, mourned and celebrated by the faithful. In a way, it continues to unroll at all times, reinforced by the presence of the icon which, itself holy, is not a mere representation of the divine event but is its transmitter, making it happen again and again whenever the icon is perceived visually and spiritually.

The reduced and simplified spatial features of an icon take away its earthly realism, and together with it, the concrete temporality of depicted events. The more laconic and simple are the visual details in the painting, the more difficult it is to tie its content to a particular real point in time. It becomes a-temporal in the sense that it may have never been present in the embodied sense, or it may be omnipresent as the idea which the depicted event is meant to only symbolize.

The above examples illustrate the dependence of reality in general and of time in particular on spatial precision, and are contrary to Bergson's attempts to downgrade the ontological significance of space either by eliminating it from being or including it into time. We reply that space is not a mere artificial construction of our mind, facilitating our comprehension of the essentially fluid and shapeless reality, but an ontologically firm and irreducible fact, and replace the Bergsonian disjunction 'time or space' with a conjunction 'time and space', in the same way as we did with qualities and relations.[17]

3   Space and Extensity

Despite refuting space in general, Bergson accepts concrete extensity, which consequently may be taken as a foundation of the Bergsonian theory of space. However, as far as we can see, such a theory would be inconsistent with the theory of pure duration, and cannot be accommodated by it. In order to account for Bergson's philosophy as a whole, we will give a brief overview of it here, though it will not be possible to pursue it further in our further discussion.

In essence, space is understood by Bergson as 'an empty homogeneous medium' (*TFW*, p. 95). A more detailed definition is this: 'Space is what enables us to distinguish a number of identical and simultaneous sensations from one another; it is thus a principle of differentiation other than that of qualitative differentiation, and consequently it is a reality with no quality' (*TFW*, p. 95). Bergson distinguishes between extensity and space, the former being an inner property of a material thing, that very portion of space that it fills, and the latter being a homogeneous medium in which material things are juxtaposed. Bergson's view on extensity fits in with his view on time and space as expressed in Table 2.

17   For more on problems associated with the understanding of space and time see Franz Brentano, *Philosophical Investigations on Space, Time and the Continuum*, transl. Barry Smith (London, New York: Croom Helm, 1988); J. J. C. Smart, ed., *Problems of Space and Time*, especially Richard Taylor, 'Spatial and Temporal Analogies and the Concept of Identity', 381–96, for a demonstration that temporal and spatial relations are 'radically alike'.

Table 2 Bergson's view on time and space

Mode of existence \ Space & Time	Space	Time
As a medium uniting all beings	✓ Exists as a homogeneous medium without quality.	✗ Does not exist as a homogeneous medium because a homogeneous medium void of any quality is space.
As part of the makeup of beings	✓ Exists as extensity of things in a heterogeneous reality.	✗ Exists as duration of beings taken as processes.

A common view on time and space is that they are both homogeneous media: in the first one, objects succeed one another and in the second, objects exist simultaneously. Also, we commonly believe that objects that exist in time and space are not outside time and space, but contain time and space within themselves. Bergson distinguishes between time and space as supposed media, and time and space occupied, appropriated, by objects: the latter are modified space and time, extensity and duration, part of the makeup of things. They appear as much a quality as, say, colour or weight.

Bergson's idea of extensity could be used to provide an alternative solution to the Eleatic paradoxes. By his solutions of the Eleatic paradoxes Bergson wants to prove that motion is real, that it has no spatial features. Also, he aligns sensory data and reality showing that it is possible to grasp the real through the senses. But more importantly, Bergson's defence of the ontological status of motion, his despatializalazation of it, clarifies his position regarding the relation between motion, consciousness and time. It is not simply that the perception of movement is duration as a psychic state; movement itself is duration. The reinforcement for this interpretation of Bergson's position can be found in the following passage, which contains his analysis of Zeno's paradoxes.

It is to this confusion between motion and the space traversed that the paradoxes of the Eleatics are due; for the interval which separates two points is infinitely divisible, and if motion consisted of parts like those of the interval itself, the interval would never be crossed. But the truth is that each of Achilles' steps is a simple indivisible act, and that, after a given number of these acts, Achilles will have passed the tortoise. The mistake of the Eleatics arises from their identification of this series of acts, each of which is of a definite kind and indivisible, with the homogeneous space which underlies them. As this space can be divided and put together again according to any law whatever, they think they are justified in reconstructing Achilles' whole movement, not with Achilles' kind of step, but with the tortoise's kind: in place of Achilles pursuing the tortoise they really put two tortoises, regulated by each other, two tortoises which agree to make the same kind of steps or simultaneous acts, so as never to catch one another. Why does Achilles outstrip the tortoise? Because each of Achilles' steps and each of the tortoise's steps are indivisible acts in so far as they are movements, and are different magnitudes in so far as they are space: so that addition will soon give a greater length for the space traversed by Achilles than is obtained by adding together the space traversed by the tortoise and the handicap with which it started. This is what Zeno leaves out of account when he reconstructs the movements of Achilles according to the same law as the movement of the tortoise, forgetting that space alone can be divided and put together again in any way we like, and thus confusing space with motion. (*TFW*, pp. 112–14)

But there is something that Bergson himself leaves out. Giving an account of an instance of movement, he admits two elements – space traversed and movement per se, duration. The moving body, the one that traverses space, *the one that moves*, is left out of this discussion. In relation to the arrow paradox,[18] together with Russell we can say that in Bersgon there is 'flight but no arrow'.[19] The moving body, considered closely, gives rise to a more complex dialectic of space and time, given Bergson's division of the spatial into a spatial medium and extensity. We can understand an object as a spatial object in as much as it is extensity. This extensity does not necessarily have to occupy a fixed position in space as a medium. Thus the connection between the 'portion of space', which is the extension of the moving body, and the place it occupies, is loosened. The same 'portion of space' appears

18   A flying arrow is, in fact, at rest. At any indivisible instant of its flight, the arrow occupies as much space as when it is at rest. Therefore, it is always at rest.
19   Russell, *The Philosophy of Bergson*, 18.

in two guises, as space divided in itself. As extensity, it shifts in relation to its previous position in space as a medium, and in relation to other objects; and movement results out of this division of space into two parts.

Bergson's division of space into a medium and extensity may provide an alternative solution to the 'arrow' paradox. Zeno says that the arrow, when it is moving, occupies the portion of space equal to itself, just as when it is at rest. When this is asserted, what is assumed is that the arrow occupies a position in the spatial medium. But, when the arrow occupies the space equal to itself, it is primarily its own internalized space that it occupies. The arrow's extensity is a property of the arrow, and as such it is intrinsic and undetachable from the arrow. Therefore it is not contradictory to the idea of movement that the arrow should occupy the portion of space equal to itself: all it means is that it remains itself whilst it is moving. However, at different times it would occupy different portions of space as a medium – and that is evidence in favour of movement, which the Eleatics wanted to disprove.

4 Three Hypostases of Time

The irreducibility of temporal qualities and relations confirms that we must talk of temporality in two guises at least: the temporality of earlier and later, and the internal temporality of events; temporality as a relation and temporality as a quality. But, if we assert temporal essences and relations between them, then we will need to assert time that unites them all, universal time. Thus, instead of the idea that there must be one and the same time, whatever its nature, we get an idea of three hypostases of time:

1. Time as duration of individual phenomena (time as a quality).
2. Time as temporal relations.
3. Universal time.

Bergson addressed time as individual duration and dismissed temporal relations and universal time as space in disguise. Asserting all three, we need to establish which hypostasis of time should be given ontological priority. Clearly, this cannot be temporal relations, for if we imagine that 'earlier' and 'later' float somewhere in suspended existence, waiting to be embodied in reality, then we inevitably substantiate these relations and turn them into qualities. But if time is first of all to be treated as a quality, should it be the temporal stretch of an individual phenomenon or the universal, empty time waiting to include temporal reality?

Problems with first admitting some kind of empty time are not easily resolved. Does empty time have a beginning? What defines its boundaries, if it has any, while the content that fills this time is still absent? What defines the direction or arrow of such time? These difficulties are insurmountable if we begin to talk of time in global terms, and it is inevitable that the discussion on time should begin with the temporality of an individual phenomenon, as in Bergson. Then an attempt can be made to derive other modes of temporal existence from an individual duration.

We can take as a starting point an observable temporal phenomenon – any phenomenon, an object or an event – inasmuch as all phenomena are temporal. But, unlike Bergson, we need to go further than just acknowledging the inner duration of things: we must derive from it temporal relations and the concept of universal time.

As discussed above, the reason why Bergson would object to including temporal relations such as 'earlier' and 'later' into the temporal system is that their acknowledgement requires a mental alignment of terms which takes place in simultaneity, thus presenting time as a static, non-progressive system, where everything is set once and for all. A Bergsonist would rightly insist that the theoretical temporal framework must be able to account for the dynamic, ever-progressing nature of temporal reality. However, by adapting the vocabulary, it is easy enough to introduce the dynamic element here. Instead of saying that terms *are* earlier and later than other terms, we could say that terms *happen* or *take place* earlier than some and later than others. For, regardless of the possible Bergsonian criticism that any comparison of terms presupposes their simultaneous alignment which defies the successive nature of temporality, events do happen earlier than some and later than others.

Now, Bergson would say, this gives us a dynamic picture of the temporal flow in between terms, but the terms themselves, those that happen earlier or later than each other, are still portrayed as fixed entities, as nodes of a stable, definite physicality, as reality that, albeit existing in time, nevertheless lives through it with its nature being unaffected by it. So our next step will be to introduce temporality into the terms of temporal relations as well, and to present them as events or processes. Bringing in the Bergsonian language, we will recall that temporality of an event constitutes part of its makeup and contributes to its unique identity. Thus the relations of earlier and later will be applied not to unchangeable nodes of reality but to entities that change in time themselves. 'Earlier' and 'later' occur between terms which themselves are never brought to a standstill; they are relations that emerge in the unceasing dynamics of change. We could say that temporal qualities, whilst unceasingly growing, transcend themselves in relations with other qualities, as well as producing the network of relations within themselves. Thus relations of earlier and later themselves partake of the dynamic nature of time.

From this perspective, we shall suggest that the qualitative temporal component of a phenomenon emanates temporal relations which the phenomenon has or may have with other phenomena. This in its turn reveals the significance of spatial definitions for temporal identity of phenomena. It would not be accurate to say that duration as a pure temporal extract from the phenomenon evokes relations of, say, earlier and later with other phenomena. An individualized duration is by no means such a pure temporal extract because, first of all, it being individualized means existence within rigid boundaries. It is impossible to say that these boundaries are exclusively temporal, say, because they are set from time t to time t', since only spatial markers can set such boundaries. Duration as one's conscious life, for example, is defined by the physical existence and coherence of one's living body. Duration as movement of a body is equally marked at the beginning, by the body starting to move and at the end, by the body stopping.

Also, if events of this type happen in a definite period of time, they also necessarily happen in a definite stretch in space. Spatial features not only mark the temporal boundaries of an individual duration, they take part in the evolution of duration throughout its continuous existence, for each duration is defined by a specific process which always has a physical backing. If the physical components of this process cease to move or, more accurately, alter their pattern of movement, then the given duration will cease to exist, or it will change.

Due to the necessary involvement of the physical component, spatially precisely located in relation to other objects, duration as a temporal component of the process in question, will be also spatially located – contrary to Bergson trying to eliminate the 'where' issue from the so-called true time. We locate psychological duration when we speak of the whereabouts of a person because all his or her thoughts and feelings take place wherever he or she is. We locate the duration of an action and of any other process because we locate the trajectory of the moving body, which has definite spatial parameters. Thus time is spatially located, in this sense.

We insist that duration is by no means a purely temporal reality, independent from spatial markers and void of spatial content, but is an entire individualized phenomenon, with all its spatial and temporal parameters. For if we talk of motion and the temporal extension of an object, we cannot abstract them from physical and spatial features without losing duration itself. This, in its turn, means that when we say that duration as a quality prolongs itself in temporal relations, the agency that emanates those relations is not the stretch of pure time, as such unembodied time cannot produce any relations, but the entire complex compound of a spatially located physical or physically backed process, event or an object.

In fact, the feature that represents objects in their temporal relations is not their own temporality, even defined by spatial features, but their visual spatiality itself. We remember that Bergson would dismiss temporal relations as spatialization of time, because the terms of such a relation would be juxtaposed in quasi-simultaneity. What really happens then is that the terms of a temporal relation appear in their guise as spatial entities, showing their visual spatial side for a comparison or contrast, necessary for forming a temporal relation. When we compare temporal relations of objects which reflect their ontological happening earlier than some and later than others, we are not interested in their inner temporality as such but in their temporal individuating boundaries which are spatially defined. That is why that which enters into a temporal relation of earlier or later is what Bergson called a crystallized object, i.e. an object or event whose spatial qualities have come to the surface as those that are most relevant for the required relation. Thus the quality that emanates temporal relations is not pure time, nor even the temporal extraction of the phenomenon, but its spatial, physical part.

Bearing in mind that a quality can form a multitude of relations and if we talk of a spatial body as a quality, then we will have to admit that amongst the multiplicity of relations that a spatial body forms with other bodies there are temporal relations. The ontology of a physical body entails that it exists and endures later, earlier and simultaneously with other bodies. From this we can see that it is the spatial marking that brings to life and to light the relational aspect of temporality. As we can see from this, time can only be revealed via space, whether in the individualization of and exposure of duration, or in the formation of temporal relations.

Regarding the interplay between temporal qualities and temporal relations, it is easy to see that whereas sometimes relational facts do not seem to matter to the qualities involved, such as linking the fact of dissolving sugar with the timing of the news programme, in many other situations, temporal relations may play a crucial role for the qualitative definition and identity of phenomena. For instance, x may overhear y confessing a crime, but testify in y's favour during the criminal investigation. What would determine whether x is a socially dangerous liar or a mistaken law-abiding citizen, is the temporal relation between x's hearing y's confession and x giving the testimony. If the confession took place earlier than x's testimony, then x is a liar; if it took place later than x's testimony, then x may well be a law-abiding citizen.

But what about universal time? As a fragment of temporal reality, each entity and each event is endorsed with relations of earlier and later, and this endorsement ontologically coincides with the very fact of their existence. Apart from being successively future, present and past, each entity fixes the temporal ordering by emanating relations of earlier and later and thus securing the structure of temporality in general. For, by saying that an entity is necessarily later than some other entities, we acknowledge that there were things happening in time prior to this entity, and by saying that an entity is earlier than some other entities, we establish a guarantee of the future world. Each new event, in virtue of it being 'earlier' implies following events, and the relation of it being 'later', safeguards the infinite historicity of the past, reminding us that before the most ancient event there must have been an event prior to that one.

It is easy to demonstrate that universal boundless time follows from individual phenomena as a logical necessity. A phenomenon – every phenomenon – entails temporal relations of earlier and later with other phenomena. This means that its existence implies the existence of others, those that preceded it and those that will follow it. The implied phenomena in their turn also harbour relations of earlier and later, so that every previous phenomenon must have had a preceding one, and every future phenomenon must have a following one, with the time line projecting itself indefinitely in both directions. Thus the existence of one single phenomenon implies other phenomena which in their turn constitute a timeline saturated with content. This gives us a timeline as a logical necessity which can be constructed out of the existence of a single phenomenon. Crudely speaking, we could say that time may have began with the hypothetical first ever event – say, a big bang – but as soon as that event took place, it created universal time, which included the indefinite past as having elapsed prior to the first event.[20]

5   The Past

Having accepted both qualities and relations into the discussion of time, we are prepared to treat parts of time, the past, present and future, in either qualitative or relational terms, or in both, as required. There is no doubt that

---

20   If we do not accept temporal relations that emanate from the individual temporal quality and constitute the past retrospectively, then we face insurmountable problems trying to account for the beginning of time and temporal reality. As Kant shows, neither can we prove sufficiently the thesis that the world has a beginning in time, nor the antithesis that the world is infinite. See I. Kant, *Critique of Pure Reason*, transl. Norman Kemp Smith (London: Macmillan, 1961), 'First Conflict of the Transcendental Ideas', 396–402.

duration must be understood as 'real time, regarded as a flux' (*CE*, p. 355) which involves indivisible movement between past, present and future. In 'The Perception of Change' in particular Bergson demonstrates that the impossibility of separating the past from the present is due to the indivisibility of change and time (*CM*, pp. 152–6). The present and the immediate future equally cannot be separated from each other, for the same reason. The criticism of prediction in *Time and Free Will* especially demonstrates that the dynamic progress depends on the immediately preceding state of reality and that is why one cannot predict future states as if they were planned in some way (TFW, pp. 183–97).[21] Thus the essence of duration rests on all the temporal dimensions taken in their indivisibility. However, *Matter and Memory*'s analysis of duration in terms of the past shows that out of all three temporal parts, the most important one for Bergson seems to be the past.[22] It is the past that is accumulated in duration, survives as memory and constitutes reality. When we say that for Bergson, time equals being, this may well be rephrased as the past equals being, inasmuch as the past is portrayed as substantial and rich in content, whereas the future is non-existent, and the main function of the present is to activate the past that already exists, and to add new content to the new past.

Before we consider the content of the embodied past, something needs to be said about the past as a specifically temporal feature of reality, if only to point out that it is not the same as the content of the past, regardless of Bergson's attempts to equate time and its content. If we compare the present, past and future as purely temporal extractions from reality, we would struggle to find criteria to distinguish them from one another. Disembodied and

---

21 As noted by Mullarkey, 'The language of *Time and Free Will* fosters images of continual transformation, of multiple succession'. John Mullarkey, 'Forget the Virtual', *Continental Philosophy Review*, Vol. 37 (2005), 476.

22 This did not pass unnoticed for other readers of Bergson: see Stephen Crocker, 'The Past is to Time What the Idea is to Thought or, What is General in the Past in General?', *Journal of the British Society for Phenomenology*, Vol. 35, No. 1 (January 2004), 42–53. Sartre goes as far as to say that Bergson considers the past in isolation from the present and fails to reconnect it to the present (Jean-Paul Sartre, *Being and Nothingness: An Essay on Phenomenological Ontology*, transl. Hazel E. Barnes (London: Methuen, 1969), 110).

abstracted from physical action, from traces of events, and from any other clues for presentness and pastness, time becomes basically a foundation for continuity; not even duration, since duration includes factual content, but a basis for duration. Such time would have neither past, present, nor future parts and neither would it have a direction of flow, because all these features are distinguishable only with the catalysing effect of physical or psychological content. It is important to emphasize that this unimaginable (because we can only imagine embodied things) time cannot exist by itself, and is only a theoretical abstraction. We must remember that if we insist on a purely temporal component in reality, distinct from everything spatial, we lose reality and time with it. So, when we talk of the parts of time here, we mean embodied stretches of time, with physical or psychological content, and with spatial or conceptual markers.

Undoubtedly, the past, as a temporal stretch saturated with all the physical and psychological processes that happened when the phases of this temporal stretch were present, is a quality. But describing a temporal period as past, we correlate it with the present, i.e. with our own present position. From this perspective, 'past' is a relational characteristic of the given temporal stretch. Also, because this relational characteristic is given to the period from the perspective of another period, and after the first period is gone, then this relation may be considered purely external and unable to affect the nature of the temporal period in question. In other words, the fact that the temporal period spoken of is past in relation to now, should play no role in determining the qualitative structure in the period that we call the past.

However, it is evident that for Bergson, the nature of the past temporal stretch is determined primarily by it being past. This means that the relation that the temporal period has or acquires with the consequent period determines its own inner nature. What the ontological status of the past is, and what the ways in which it is a quality and a relation are need to be clarified.

According to Bergson, all of the past survives in a modified, condensed form. However, a stricter approach to the correspondence of the past to the present would reveal that, although all of the embodied past may indeed contribute to the formation of the embodied present, the very fact that the past survives in a modified form means that the content of the past, taken when it was present itself, is not identical with that part of the present content which is supposed to have been inherited from the past. There is some part of the embodied past that falls off and disappears with time. For instance, a tiger passes on to its descendents the tendency to have striped coats, but its own stripes vanish when the tiger dies and its coat disintegrates. So there is the past that vanishes, and it is the ontology of this past that we want to examine.

The ambivalent nature of the past that is gone consists of two conflicting facts about it. Firstly, past entities and processes are those that once were present, occupied space and acted on one another. The fact that x has once existed in the present seems to secure x's ontological status as a real thing for eternity to come. By contrast, if y had never existed, y's ontological status is obviously far less secured than that of x. But if x is past, it is nowhere to be found, just like y, as present events fill the world with their own being, leaving no room for events gone. At first glance, both x and y are equalized in the way that both are nonexistent. Judged from the viewpoint of the present time, since there is no evidence of x existing, a present time observer cannot say anything about it, as if it is indeed something that had never been. But the truth is that it has been, even though it is now disconnected from the present.

Just like the past that survives with the present evidence of it, the past that disappears has existed, unlike events which have never happened. Bearing in mind that past entities are really ex-present entities, we must ask, what happens when a present entity becomes past? Firstly, the relation of pastness is being established between the entity that has gone and all entities and processes that are present. From the point of view of the current situation, the status of the past entity is measured against its relevance to the present. The past entity does not exist, but not in the same way as something that had never existed. It existed, despite the fact that there may not be any evidence left of it in the present.

This should demonstrate that it is wrong to test the ontological status of things and events by measuring them against the background of present things and events. This is what Bergson tries to avoid by introducing the past that survives as memory, which is not necessarily accessible by our mind, but this attempt is not entirely successful because his argument that the past exists is too tightly linked with the argument that the past survives as a foundation of the present, and also because the very idea of the surviving past takes it into the present and correlates it with all things present. The Bergsonian surviving past then is simply some kind of a disembodied present![23]

From the point of view of the present, the time period that is past is, in fact, non-existent, being pushed out of the way by the present period itself, or even by other time periods that preceded the present and which now are past as well. If we relied on the present to evaluate the ontology of this vanished past, we would observe that the time period being past is non-existent in relation to the present, and conclude that it is past and non-existent universally (because we believe the presentness to be universal for all possible parallel worlds) and in relation to itself as well.

If we admit that there are things that have been and gone, then the present, that knows nothing about them, cannot have the authority to judge their ontological status. For instance, there may be events in the past which happened on distant worlds and cannot have any relevance to our present earthly existence, but this does not mean that we, who do not know about them, have the authority to say whether they existed or not.

The being, we insist, does not necessarily equal 'the present' nor is it 'relevant to the present'. If we liberate being from the necessity to be demonstrable in the present time, then we are able to talk of being outside the present and outside our scope of memory and perception. Despite the fact that some past thing may be undetectable in the present, we would still have to assert its being, not the being that *is* but the being that *was*.

Getting away from the present-centred ontology and giving up our attempts to prove the being of the past in terms of the present, we must refer to the past itself as a source of such proof. If the being of present things is asserted via their relation of presentness to themselves, what about the being of past things? Whereas they are past in relation to the present, they

---

23   At this point I may even agree with Russell's criticism regarding Bergson confusing the past and the present. See Chapter 3, footnote 12 on page 48.

cannot be past in relation to themselves. But neither are they present in relation to themselves, because being present in relation to itself requires being present universally. As we cannot say that they *are* present in relation to themselves, we can nevertheless say that they *were* present in relation to themselves and in general. If we want to talk of the past as being, we must always say 'was' or 'were' about it, because as soon as we say 'is', we correlate it with the present and render it deficient or non-existent. So, in the present-independent universe, the past being can be defined quite simply as the being that was, and this should be sufficient to equate it with the being that is, not in the least because the being that *is*, is also going to be the being that *was*.

Questioning the ontological status of the past, we talk of the events that must have been present at some time and, therefore, actually happened. When an event is present and is happening, we can say that it is also present in relation to itself. This may seem like an unnecessary tautology, but being present to itself is a necessary feature of actuality, and being present to oneself is a necessary feature of a living consciousness. So when the present event gives way to a new event and becomes past, it being past means that it is past in relation to the new event, but in relation to itself it can only be present, inasmuch as everything existing or imaginable is contemporary with itself. An attempt to incorporate pastness into the nature of an event turns out to be futile. We inevitably have to admit that whereas presentness involves a relation between oneself and other contemporary processes and between oneself and oneself, pastness cannot be a relation of oneself to oneself, but only between previous and consequent events. Thus an event cannot really be past all by itself and comprise some specific past inner nature. Although, following Bergson, we treat an embodied past stretch of time as a quality, we renounce the idea of pastness as a quality, and accept it only as a relation. Saying that the past is a quality, we mean that its eventual content is a quality, but by no means that its pastness is a quality.

6   The Future

If we used the present tense standpoint as the ultimate platform of ontology, the reality of future things would be even more questionable than the reality of past things.[24] If, from the point of view of the present, past events do not exist but had existed in some other present, future things are as ontologically inadequate as things that had never existed, because future things, indeed, have never existed. However, it is possible to consider future events as a self-projection forward in time of the present, and then the future is not non-existent in the same way as something that cannot exist, because its existence is expected and is, to some extent, predictable.

However we would find it difficult to establish qualitative features of future events. Qualities, whether accessible by mind and senses or not, must have some definite form: for example, if a thing is round, it is definitely round, and not square or triangular at the same time. But this is not so with the qualities of future events and future entities. States and events, consequent in relation to the present, are inevitable (because it is inevitable that there will be a future) but, prior to their actual embodiment, their exact nature and exact parameters are uncertain. Agreeing with Bergson that the future states of the Universe are not predetermined and that each new state is original in principle, we deny that the future states have any inner structure and any inner qualities. Due to the fuzziness of its indefinite parameters, we cannot establish the tautology of the future content and hence ascribe identity to the future which, vague as it is, cannot have a relation of presentness with itself.

As the future is uncertain, its projection consists of a numerous variety of alternative possibilities that are all imagined to occupy the same place and time in the future. The insurmountable difficulty in determining future

24   Considerations of the future are virtually ignored in Bergson, as Cunningham points out (Cunningham, *A Study in the Philosophy of Bergson*, 118). My analysis of the future in Bergsonism is based on what Bergson would have said rather than on what he actually said.

qualities will lie in this haziness and mutual overlapping of alternative properties. For when thinking of the future as a range of possibilities, we ascribe to the same future period of time different incompatible characteristics, and therefore cannot fix and identify future qualities. To give an example, the future of a sprouting seed entails a range of possible states. Only one of these possibilities will actualize itself, but while it has not actualized itself, we may think of the whole range of possibilities as the plant's future. We can imagine, for example, that at the same time t and in the same place n there will be 1) a rose bush with seven flowers; 2) a rose bush with five buds and three flowers; 3) a rose bush with no flowers, etc.

One can accept futurity as a relation between the embodied present and consequent states and events which are not yet defined and embodied. The only definite characteristic of the future that is possible is the relation of futurity between the future and the present, established in the present, and the only definite fact we can establish about the future, whose content lacks identity, is that it is future.

Thus the only indication of the being and of the reality of the future comes from outside the future, prior to its own existence, as an external relation. Thus we have a situation where one term, the present, has a relation of futurity with another term, which is indefinable, having neither identity nor inner structure. Should we admit here that the relation of futurity is that element that constitutes any idea of the future at all? If we do that, we can end up asserting a relation prior to one of the terms and fall into substantiating this relation and presenting it as a quality instead of letting it appear as a side effect of the mutual co-existence of two terms.

Actually, this substantiation occurs in our practical attitude to the future, primarily realized in relational terms, which have a qualitative value for us. We know that, whatever the content of future events will be, the relational framework for them is fixed: whatever happens, Christmas day will fall on the predicted day, and if today is Monday, tomorrow will be Tuesday. We treat our own lives lying ahead, and those of others, like an empty diary, which guarantees relational facts such as dates and ages, which will definitely become present regardless of their qualitative content.

In the ontology where the present is not the measure of being, the view of the future from the future position can be as follows. In order to establish the relation of tautology of the future event to itself, it needs to shape itself. This shaping has not happened yet in the present, but it *will* happen, so all we have to do and all we can do is wait. In time, future possibilities define themselves whilst becoming present, and then we can say retrospectively that this present was the future of that other present which is now past. So, in the present-free ontology, future events enjoy the same status as past and present ones, because they will define themselves in their own time, and for the present-free ontology it does not matter when being is present – it is real as long as it was, is, or will be present at some point. As for the reality of future things as future, just as we could not identify the past as a quality, in the same way there is no quality called futurity, as there is nothing whose inner nature would be characterized by pastness or futurity.

7   The Present

There is an unspoken assumption that real things are present, and we tend to think of past and future things as somehow less real than present ones. From the point of view of this present centred ontology, the present appears to be in a privileged situation compared to the other two parts of time: a present process, apart from being present to itself, is also present to other processes, and so its reality is secured both from within and from outside.[25] One may say that the same could be true of past and future processes as well, namely that they are present to other past and future processes respectively. But whereas past and future events are not capable of interaction and mutual acknowledgement, present events, in virtue of their being co-existent with each other in the present, realize this co-existence via an actual, physical or psychological impact on one another as they unroll.

---

25   It is from this position that Augustine gives the present a privileged status, even though he is unable to explain the present (St Augustine, *Confessions*, 267).

For instance, if a living consciousness were to forget that it exists in the present whilst daydreaming, some external intrusion, such as a ringing telephone, will inevitably bring it back to reality, i.e. to the realization that it finds itself right at the heart of the present situation. The situation, with its stimuli that require responses from us, will also convince us that the present in which we are and which we constitute, is not merely a subjective mode of existing and perceiving but an objective present, whose demands we cannot escape.

As complex processes, capable of self-acknowledgement and of acknowledgement of other events, we are sure of our present existence, and of the existence of the present time. The present is self-evident and concurrent with us; we neither leave it behind, nor stay behind it ourselves. What is 'now' and what is 'present' is obvious to us on a pre-conceptual, practical level. Our understanding of the present is effected via the acknowledgement of tautology of all present entities and processes, where everything is unambiguously definite, and is undoubtedly itself in its fully embodied form. Since entities and processes that we observe in the present demonstrate unambiguously finalized and definite parameters, we can say that the embodied present may be comprehended in qualitative terms; that the present in itself may be a quality.

The nature of the present, self-evident as a quality and via the relation of tautology, becomes problematic when we try to extract it from the immediacy of a conscious understanding of it and try to align it with other parts of time, the past and the future, in order to establish the purely objective parameters of the present. Bergson avoids this problem by refusing to objectify time altogether and labelling the objectification of time to be a substitution of space for time. Thus the Bergsonian thought, albeit giving preferential treatment to the past, does not escape from the immediate conscious apprehension of the present. For Bergson, if time as the accumulated reality or duration could be reduced to the past time, intuition, which is the recognition of such duration, could be equated with the present time in which the accumulation of reality and recognition of this accumulation takes place. The very same thing can be said of the intuited duration: if, as we remember, duration can really exist as apprehended from within, it must be a currently living, existing duration, with consciousness that can account for events which are looked forward to. However this acknowledgment from within is possible only in the present, which firstly, identifies duration with the continuous living present and secondly, limits the Bergsonian account of time to such a present.

But what are the purely objective parameters of the present? What we call present entities or processes are processes that run concurrent with each other and are able to interact with one another. From this perspective, presentness is a relation of mutual simultaneity shared by concurrent, active processes. One of these processes is our consciousness, which labels all processes concurrent with itself, present. The inevitable participation of consciousness in the ostensive naming of the present raises suspicions regarding the authenticity of the present and its objective existence, independent from consciousness. If we eliminated consciousness, would there still be present processes in the same sense? If the present is a subjective definition, would not the consciousness of a different type see presentness where we see pastness or futurity?[26]

But our immediate apprehension of the present presupposes a process that has partly been already completed and partly not yet completed, thus our present is continuous, and partakes of the past and the future. The involvement of the past and the future is so strong, that, when trying to align the objective present with the objective past and the objective future, the continuous present breaks down into the past and the future, and it is impossible to pinpoint the actual present, that very moment of transition between the past and the future, where the past is not yet past, and the future, not future any more.[27]

---

26   There is an opinion that tensed view of reality is subjective and does not reflect the objective state of affairs. Tenseless theorists propose to eliminate tenses from the discourse of time and accept only relations of earlier and later. For a consideration of both sides of the argument see Sartre, Jean-Paul, *Being and Nothingness: An Essay on Phenomenological Ontology*, transl. Hazel E. Barnes (London: Methuen, 1969).

27   Whereas Augustine regards this as a problem for the identification of the present, Loizou asserts the continuity of the present: '[I]f we followed Augustine in eliminating the continuous present tense, thus excluding extended events from ever being legitimately spoken of as present, we would thereby destroy the present tense in any form' (Loizou, *The Reality of Time*, 41–2).

However, although we fail to point at the present ostensibly, whilst it is present, we know where and when we should have detected it. In this we are guided by the decisive feature of presentness: the flowing present necessarily culminates in the physical embodiment of action and the physical contact with other present entities. The key feature of the presentness, as supplied by Bergson, is not that present entities are consciously acknowledged, but that present processes are active. When they interact, they affect each other, and we can witness the emerging changes as they unroll. Neither past nor future entities are capable of inflicting change on one another or on themselves in the same way as present ones do. It is conceivable that these interactions with consequences, which characterize present reality, take place irrespectively of the monitoring involvement of our consciousness. As our consciousness, solely by virtue of its observing powers, does not change the fact that the worldly activity takes place, it must be equally irrelevant to the fact that this activity is present. We assume that the present unrolls by itself and that the waking consciousness is able to witness it because it is itself present and active.

The question then arises: for how long do the unsupervized processes remain present? For with the involvement of consciousness, we can talk of extended present processes, those that have begun but not yet ended. In the search for the objectivity and length of presentness, it may be useful to refer again to influential activeness as its criterion. Accepting that influential embodied activity can only take place in the present, we would also have to admit that wherever we see evidence of change, creation, and interaction of processes, we must assume that this world-shaping activity must have taken place during some present, with or without conscious involvement. As for pinning down the present, again appealing to the fact that all interactions take place in the present, we can imagine two interacting objects, such as a pen and paper, and say that the present time of the writing process corresponds to the phase of physical interaction of the pen and the paper, where the pen touches the paper whilst sliding on its surface, and the paper resists the pressure, nevertheless showing the ink traces on its surface. However impossible it may be to capture the actual phase of interaction, it does take place and it must constitute the objective present time. Unfortunately, consciousness impedes the perception of the present time, always involving the immediate past and immediate future in its perceptions, because an act of consciousness, perception or thinking, is in itself of some length, so that the act of consciousness, in an attempt to grasp the present will never temporally correspond to the time of action that constitutes the present: no matter how quick our mind tries to be in grasping the present, action will always be quicker than the act of consciousness that tries to capture it.

By denying that consciousness plays a role in constituting present-ness, we assert that present processes are not present only in relation to some other processes, but are present universally and absolutely, and that all active processes in the universe must run concurrent with any other imaginable active processes, wherever they are. So, whereas the presentness is a relation of simultaneity between concurrent active processes, it must also be regarded as a quality for the following reasons. Firstly, presentness is involved in the effective activity of processes, and it is the key feature associated with its effectivity. Secondly, presentness as a mode of existence is absolute across all possible worlds. Thirdly, to be present to itself means to be present in relation to every other present thing, and when an entity is no longer present in relation to the present world, it cannot be present to itself either, and it is at this point that presentness partakes of both relation and quality. Of relation, because a present process is present in relation to other present processes, and of quality, because a present process is such in its own right, absolutely. A thing cannot be present to itself unless it is uni-versally present, because if it is not present absolutely, its ontological status is not firm enough to establish a relation of positive tautology with itself.

The present time of one present entity coincides with the present time of all other present entities. The present pierces through the entire community of worldly working phenomena, and it is this universality that raises the question of the possibility of its independent and authoritative existence, leading to a temptation to regard the present as a criterion of reality. We accept that the presentness is a criterion of reality, but we refuse to treat any particular present time as such. Whatever phenomena that have existed and contributed towards worldly development are not limited to the concrete present time. All world-changing phenomena have been, or will be, active in the present, but not necessarily in one and the same present. There have been and there will be innumerable present moments with innumerable active processes.

8   Continuity of Presents

When we align all three parts of time we see, first of all, where the main difficulties of the tenses lie. Whereas a philosopher would naturally treat them as objects of thought of the same order, our analysis above reveals that they are not. When we say, following Bergson, that embodied time is a quality, and attempt to treat its parts in qualitative terms, then we arrive at absurdities.

If we say that the past is a quality, we imagine that pastness ought to belong to the past stretch of time as its intrinsic property, but it can only be past in relation to other following periods of time, and cannot be in a tautological relation with itself. It cannot be past by itself, because this would require an inner split, as the past would both be and not be: as past in itself, it would have expired in relation to itself. Taken as a real quality, the past must be present to itself, i.e. must have preserved its integrity. Thus the past, when taken in itself, turns out to be neither past, because it cannot be gone in relation to itself, nor present because it is not present but past.

The future cannot be a quality since its characteristics are not certain until the future becomes actualized as the present. In relational terms, the future is a relation established in the present between the present and no second definite term. Only the present, when taken as a quality, is feasible as continuity of actual experience. However, considered in relation with the past and the future, it disappears, divided between them.

These absurdities reveal that since pastness and futurity are not intrinsic properties of temporal reality, it cannot be past or future in itself, as a quality. Pastness and futurity are relations between terms that appear in a temporal sequence. The present, on the other hand, is a quality inasmuch as everything existing, when present, is present in itself and not merely in relation to some other terms. Rejecting the view that reality should be centred on one particular time, such as the current present, we suggest that all reality is present both to itself and absolutely at some point, and that the idea of a reality that is only past or future is self-contradictory. To accommodate all temporal reality, we have to give up any attempt to standardize the tenses, as all three clearly have a different ontological meaning, and accept the idea of a continuity of present times, replacing one another.

Thus we arrive at the idea of the multiplicity of presents, which is not unlike Bergson's own. As Mullarkey observes, Bergson is able to talk of 'present without supposing any *particular* presence as metaphysically unique or normative'.[28] Indeed, this follows naturally from Bergson's asserting the ontological validity of the past.

28   Mullarkey, 'Forget the Virtual', 477.

CHAPTER 7

Duration: From General Idea to Concrete Self

1 Concretizing Duration as Self

This chapter will offer a transition from duration as the all-embracing metaphysical term to duration in its concrete existence. Whilst exploring time and temporal reality, Bergson arrived at the idea of duration which he presents as the most fundamental principle of being in a general sense, without, however, offering a systematic exemplification of duration in the world.[1] He indicated its various hypostases as movement, matter, life, self, universe, but did not develop a complete and exhaustive picture of any of them.

Bergson's accounts of various types of duration make us assume that within Bergsonism we can ascertain the continuity of everything that has ever existed. This continuity is based on the implicit assumption that the universe and everything in it involve movement of some kind and thus the term duration is applicable to any type of being: the same movement, albeit with different rhythms, flows through different media, and therefore everything – inanimate and animate matter, consciousness – is ultimately of the same nature and forms a universal continuity.

But the sameness of everything can only be asserted at the metaphysical level, because concrete manifestations of being are diverse. The difference

---

1   Whereas Keith Ansell Pearson, following Deleuze, tries to make sense of duration as abstract notion and a concrete phenomenon at the same time, I separate duration as a metaphysical principle from its concrete manifestations. See Keith Ansell Pearson, *Philosophy and the Adventure of the Virtual: Bergson and the Time of Life* (London: Routledge, 2002), 38–42.

between various types of being is characterized by the rhythms of their durations, so in Bergsonism we see a hierarchy of being, where lower levels of being are characterized by slower rhythms, and higher levels by faster rhythms: from inert material objects at the bottom to human beings at the top, and biological life in the middle. Inert material objects demonstrate internal duration on the micro level and displacement on the macro level. Living organisms combine material and biological processes. Human beings exist in all three levels and comprise the full hierarchy of being as material objects, as living bodies and as psychological duration.

Whilst scrutinizing the idea of heterogeneity in Chapter 5, I found just how significant the whole is when it comes to characterizing the nature of parts. Since heterogeneous duration is temporal, the whole of it is synonymous with 'later', and part is synonymous with 'earlier', because the whole of duration comprises its total history which, in temporal terms, comes later than any of its parts. Also, as duration gains new content over time, we can equate 'earlier' with 'simpler' and 'later' with 'more complex', and say that since 'earlier' is part of 'later', then 'simpler' is part of 'more complex'. We can also say that duration in its up to date state, as an all-embracing concrete duration, as a heterogeneity with highest possible complexity, is duration taken at its latest – in the present.

The entirety of being taken in the present includes the history of the universe crowned by the current human existence. In this chapter we will demonstrate that if we want to exemplify and concretize heterogeneous duration as embracing everything that exists, we must present duration as consciousness, because only consciousness can account for other types of being, such as biological and material existence, and other types of being cannot reach out beyond themselves and explicate conscious existence. (For Bergson, of course, consciousness is not necessarily tied to human form, and in *Creative Evolution* it is universal (*CE*, p. 220).

The concept of duration has its own limitations, however. Bergson believes that sharing the same characteristic of duration is a basis good enough for the explanation of the union of spirit and matter. Indeed, mind is duration of psychic states, and body is both material duration with its vibrations and displacement and a living process with growth, inherited tendencies and actions. They both comprise memory which in the mind binds psychic states into one continuity, and in the body preserves tendencies, extracted from past events. Both mind and body grow: mind via perception, recognition and volition, and body, via organic growth and cell renewal.

It seems as though Bergson wants us to regard life as a prolongation of material movement, and fully developed alert consciousness as a prolongation of life. Biological processes, according to Bergson, are conscious or quasi-conscious inasmuch as they demonstrate inner logic, coherence and enclosure of the previous development into a present act. From this position, human consciousness is but a further stage of this complex combination of movements. Bergson claims that the relation between mind and body 'must be established in terms not of space but of time' (*CE*, p. 220), which means that they should be considered as processes prolonging themselves into one another.[2]

Indeed, we will see this continuity in perception (see our analysis of the feeling of warmth below). However, demonstrating the unity of mind and perceived matter does not explain the bond of mind and brain, and this is where the potential of duration is exhausted. Less complex forms of being are unable to demonstrate predisposition for more complex forms of being until they are already part of that being, and matter by itself is unable to show the potentiality of the union with consciousness. The best we can do is indicate components whose relation to each other may be regarded as a proto-relation of mind and matter. Those components are: in inanimate matter, *movement* and *substance*, and in organic matter, *living force* and *matter*. Movement in inanimate matter comes in two guises: as *movement of subatomic particles*, whose permanence of rate maintains the chemical identity of the material object, and as *displacement*, the movement whose parameters are not as rigidly defined but which, nevertheless, is a necessary component of the universe as a whole. As a prototype for the

2   Karin Stephen holds the view that in Bergson, matter and mind appear as 'incompatible abstractions' whereas they are 'blended' in 'the actual fact' (Karin Stephen, *Misuse of Mind: A Study of Bergson's Attack on Intellectualism* (London: Routledge, 2001), 81).

life of consciousness, *internal vibrations* could be compared to the *internal living process* that maintains the organic existence of the individual, and *displacement* with *action*. In organic matter, we find three types of movement: *evolutionary movement, the living process* of an individual organism and the *behaviour* of an individual organism.

2   Imperfect Duration in Inanimate Matter

When in Chapter 5 we discussed the whole in temporal terms, as comprising later and earlier parts, we also established that the nature of the earlier parts becomes affected by the later parts: later stages of an event comprise longer history of that event, which includes earlier parts, retrospectively enriched by relations with consecutive events. So if we want to examine the up to date status of a past event, or a state of affairs, we should take into account everything that followed it as well, because what followed would have altered the nature of that event or state. Applying this to worldly development, we will observe that earlier states of the world, for example when it comprises only inanimate existence, cannot offer any indication for being at later states of the world which comprise life and consciousness, nor can it offer clues for the role of matter in biological or conscious existence. On the other hand, the state of the world comprising life and conscious existence can account for inanimate matter as a foundation for life and consciousness and as images that consciousness accesses by way of perception. But this relation of life, consciousness and matter only appears when consciousness and life are already on the scene. Prior to the existence of life and consciousness, matter in itself does not contain any predispositions for its possible relation with biological and conscious structures – this relation is established retrospectively. Duration, taken at the level of matter, is limited to matter and is unable to explain anything beyond inanimate matter.

As we remember, consciousness, for Bergson, is a movement tending towards a higher intensity, whereas 'matter is a relaxation of the inextensive into the extensive' (*CE*, p. 230). The movement of life and consciousness is 'ascent'; the movement of matter is 'descent' (*CE*, p. 12). The crucial point is that both matter and consciousness are seen as ultimately the same movement which has split into two opposing tendencies: Bergson talks of the real that passes 'from tension to extension and from freedom to mechanical necessity by way of inversion' (*CE*, pp. 249–50). It is as if the same movement that defines the type of being is alterable, and can constitute either matter or spirit, the words 'the same' here playing a decisive role because they indicate an ontological transitory link between matter and spirit, making them part of the same process reality in principle: 'life is a movement, materiality is the inverse movement' (*CE*, p. 263).

Bergson 'affirms the reality of spirit and the reality of matter' (*MM*, p. 9) and stresses that they are the streams into which the being divided itself, and one is by no means a derivative of the other. He treats matter and spirit as polar qualities, as building blocks of the universe, equal to one another in their temporal existence and equally participating in concrete manifestations of heterogeneous being, which includes life and consciousness.

We argue that matter and spirit cannot be taken as ontologically equal polar qualities because matter is a lower, less complex being than consciousness, and can be comprehended as an element of a higher being of life and consciousness only from the position of this higher being. And as we equate 'more complex' with 'later', we can say that matter, as an ontologically simpler, less developed and in this sense 'earlier' type of being, forms a relation with the more complex, more developed, 'later' type of being which includes matter as its foundation, only when it is already part of that higher being and not before, as this relation emanates from the higher being to matter retrospectively. Therefore we cannot follow Bergson in equating matter and spirit and giving them equal roles in the union of matter and consciousness: if there is transitory movement between them, it originates from spirit and not matter.

Although duration involved in matter cannot progress beyond itself and embrace the totality of being, this is still a valid case of duration for Bergson. Taking matter in its pre-animated state, we remember that at first it appears in the Bergsonian philosophy as a purely spatial and static mix, with no history, as an ever-renewed present with no duration and no provisions for genuine change. Later we realize that this a-temporal picture of matter is nothing but an ideal picture of materiality, as real materiality turns out to be far more complex.

Bergson invites us to take the data of science into account and imagine the ultimate structure of material objects which, free from the view we have of them, lose their definite shape and their apparent stability. 'We see force more and more materialized, the atom more and more idealized, the two terms converging toward a common limit and the universe thus recovering its continuity. We may still speak of atoms; the atom may even retain its individuality for our mind which isolates it, but the solidity and the inertia of the atom dissolve either into movements or into lines of force whose reciprocal solidarity brings back to us universal continuity' (*MM*, p. 200). And more: 'The nearer we draw to the ultimate elements of matter the better we note the vanishing of that discontinuity which our senses perceived on the surface' (*MM*, p. 201).

As our habitual view of the material world turns out to be a practically useful illusion, we find that in matter, taken in itself, discontinuity of definite shapes dissolves into a continuity and cohesion on the atomic level. The solidity disappears too. On the microscopic level we are shown 'pervading concrete extensity, *modifications, perturbations*, changes of *tension* or of *energy* and nothing else' (*MM*, p. 201). Bergson believes he has proven successfully that consciousness is constituted by the inner movement of psychic states in *Time and Free Will*, and in *Matter and Memory* he attempts to present matter in the same fashion. He wants to show that particular qualities of matter are in themselves constituted exclusively by movements with particular rhythms: 'Matter thus resolves itself into numberless vibrations, all linked together in uninterrupted continuity, all bound up with each other, and travelling in every direction like shivers through an immense body' (*MM*, p. 208).

When one talks of movement and vibrations, there is a need to clarify what moves and what vibrates. A sympathetic reader of *Matter and Memory* may suggest that, even though Bergson seems to disregard that which moves, he nevertheless does not deny its existence. However, *The Creative Mind* contains a passage which explicitly affirms the opposite. 'There are changes, but there are underneath the change no things that change: change has no need of a support. There are movements, but there is no inert or invariable object which moves: movement does not imply a mobile' (*CM*, p. 147).

Appealing to the auditory experience of perceiving music, Bergson claims that if we eliminate from it spatial references such as images of notes written on a piece of paper, the keyboard and the musicians, then 'pure change remains, sufficient unto itself, in no way divided, in no way attached to a "thing" which changes' (*CM*, p. 148). Music is only one example of substance free motion, as Bergson asserts that when we think of a moving object, '[t]his alleged movement of a thing is in reality only a movement of movements' (*CM*, p. 148).

If both consciousness and matter are nothing but pure movement, then they simply become equated and the gap between them, as in brain and consciousness, can be called an illusion of spatialization. It seems that Bergson implies just that, but his plan does not work. Lacey, clarifying the meaning of the music example, says that usually, movement, when observed, implies a moving object which retains the same form but changes its position. From this perspective, movement does not seem to affect its essence. As for the melody, he admits that without the variation it would not be itself at all. Lacey says that, for Bergson, 'the fact that change can be essential to something is not confined to things like melodies but can apply to things ordinarily accepted as enduring substances'.[3]

However, Lacey continues, even though Bergson may have succeeded in demonstrating that change is essential to things, he did not manage to prove that we deal with pure change. For if, for example, melody equals 'pure change', what would it be? 'How would one tell one change from another, or one type from another? How could a melody have structure if it had no content?'[4]

Lacey refuses to talk of sounds as void of any substantiality: 'A sound that is only accessible to the sense of hearing is in a way "insubstantial",

3   Lacey, *Bergson*, 97.
4   Lacey, *Bergson*, 99.

though it can still be objective, or at any rate intersubjective; it can be such that anyone suitably placed could hear it. It is also substantial enough to carry properties. It can be high or loud at one moment and low and soft at another, and would change from one to the other in the sense that it existed as a whole at each moment'.[5] As for the example of colour, which, as Bergson demonstrates, resolves itself into a series of vibrations, Lacey reminds the reader that Bergson talks of a coloured *spot*, and a spot, even more than a sound, has features of substantiality and objectivity.

Lacey moves on to suggest that what Bergson really does is to challenge a tradition in philosophy that regards the substance of that which is real as permanent, that no matter what changes, there must always be something that does not. Lacey's own example of something that 'changes its matter whenever it moves' and 'moves by changing it' is a water wave, which 'does have a material substrate even if this is always changing'.[6] If we look beyond the individual wave, we will find the whole of the sea changing qualitatively all the time. The next step would be to extend this idea to all motion in the universe, whose parts are constantly changing in this way. Thus, Lacey believes, we do not get rid of the mobile in favour of pure movement; we rather get rid of the permanent substrate. Yet the substrate remains, because even though Lacey agrees that solidity does not have to be a criterion of objectivity, he argues that 'properties, including movement, cannot exist on their own. Even a coloured spot is a coloured spot, with a shape, size and location'.[7]

Lacey discovers passages in Bergson where he talks as if he does accept substantiality of matter: 'There do not exist *things* made, but only things in the making' (*CM*, p. 188). Lacey says, 'This suggests he is allowing that there are things but insisting that they are perpetually changing'.[8] Bergson even asserts substantiality, according to Lacey, but it is the substantiality of change: for Bergson, 'it is action and movements rather than objects

5   Lacey, *Bergson*, 100.
6   Lacey, *Bergson*, 101.
7   Lacey, *Bergson*, 105.
8   Lacey, *Bergson*, 106.

that are the substances',[9] and thus we are invited to accept, in Bergson, 'an ontology of actions and movements',[10] and we should talk of an ever-changing substance.

Having considered Lacey's arguments, perhaps we could replace Bergson's apparent anti-substantialism with a thesis that substance exists but is never at rest. Density and motion thus constitute one and the same reality of matter, and the universal movement of movements is not a void (and it would have to be a void if we remove any substantiality from it), but a process where the element of volume and density and the element of motion are both necessary predicates of material reality, and neither one nor the other can be abstracted from it without destroying the integrity of matter. Neither can substance be imagined still, a-temporal and motionless, nor can motion be imagined empty of volume, but both components are inseparable from each other in the space-time continuum.

Bergson says that in movement there is nothing solid and stable that moves, and we accept this as a claim that it is a mistake to imagine some substance that remains unchanged whilst movement, or change, are super-imposed on it as a relation between itself and other objects, or between itself in the past and itself in the present, with the core of that thing preserved in order to carry its indestructible identity through time. This, within the Bergsonian understanding of things, portrays that a supposed indestruct-ible core of the thing would have to be regarded as a-temporal whereas nothing is a-temporal and everything is involved in time and everything changes, even if only by becoming older.

The question still remains: Bergson wanted to demonstrate that the movement of matter and the movement of consciousness are essentially the same – and they must belong to the same order of reality in order to be part of the same universal movement – but does he succeed in proving this? Our answer is no. The supposed splitting of being into consciousness and matter is not something that had occurred at the outset of the development of the universe. 'The universe is not made, but is being made continually'

---

9   Lacey, *Bergson*, 107.
10   Lacey, *Bergson*, 108.

(*CE*, p. 255): therefore, we should expect to see the same ontological processes now as that of billions of years ago. Matter must be governed by the same key laws as it was in the remote past prior to the emergence of life on Earth; it must be ready to be a foundation of life again if the sudden disappearance of life required it. So, if Bergson were correct, we should be able to see some evidence, some traces of the 'sameness' of the movement involved in matter, life and consciousness as they are now, and not just trace this 'sameness' to the hypothetical remote past where amorphous being divided itself into matter and consciousness once and for all.

The type of movement that we observe in matter is, above all, inseparably bound with substance. This bond is such that all of the movement is a property of all of the substance and vice versa, so that neither one nor the other can be considered separately. In matter, movement is in-built into the structure of things in such a way that the substantiality of the real is a process, which comprises substance and movement as part of one thing. The key feature is that movement cannot be extracted from the material object: the movement is totally committed to the physical substance of matter. Movement and substance cannot be considered separately, because as soon as matter loses one of them, it loses its integrity as matter, so that both movement and substance, taken separately, are nothing but abstractions.

Bergson's assumption that movement is the key to the connection between matter and mind is based on the observation that movement of one entity can initiate movement in another entity: the foot kicks the football, and the ball flies. For Bergson, we understand, this is a prolongation of the same movement which finds embodiment firstly, in the kick of the foot and secondly, in the flight of the football. The movement is the same because of the uninterrupted continuity of the first phase flowing into the second phase – there is no temporal or spatial gap between them. The movement is one, but it is modified in each case due to the different media involved.

One moving entity (the foot) causes another entity (the ball) to move, by kicking it forcefully. The movement of the first and the movement of the second are stages of the same movement which began with the raising of the foot and continued with the rolling of the ball, as if movement were some tendency that sought embodiment in various portions of matter. But any portion of matter, prior to any displacement communicated to it (outside movement), already comprises movement within its own structure (internal movement).[11] Moreover, a particular movement of its subatomic particles determines the type of matter we deal with: stone, gold, plastic etc. The mechanical movement from outside that causes the ball to roll or fly does not interfere with the internal movement of its atoms and electrons but remains superficial to them. Both entities, the foot and the ball, retain their integrity, and the movement that they take part in, if it is a bond, is weak and can be easily cut out with no consequence for the entities involved.

Outside movement that affects the internal movement of atoms and electrons is possible, and occurs in a chemical reaction. But in both physics and chemistry we witness material things making changes in other material things. We never see movement detaching itself from a particular substance and either transferring itself into another substance or acquiring some independent existence by itself. Besides, in mechanical and chemical changes it is not pure movement embodied in physical things that causes pure movement in other physical things: it is the whole moving substance which constitutes a particular type of matter that causes changes. That which changes is not pure movement either but a specific compound of substance and movement which constitutes concrete matter with a specific chemical structure.

Bergson argues that real movement is a qualitative change rather than simple spatial displacement. But when we talk of qualitative changes, these must be changes of qualities and, if we deal with material things, why should we exclude substantiality from a qualitative change? Instead, we should regard the whole of matter as the cause and recipient of change, and not attempt an extraction of pure movement from physical reality.

---

11   Terms 'outside movement' and 'internal movement' referring to the mechanical displacement of an entity and to atomic movements within the entity respectively are not used by Bergson, but they reflect the significant ontological distinction made in accordance with the Bergsonian principles. I claim that whereas internal movement is crucial for the identity of the entity as a particular type of matter, outside movement is not.

To summarize, Bergson is opposed to the view in philosophy where solid stuff is regarded as primary reality and movement is reduced to a superimposed characteristic. He strives hard to raise the importance of motion in ontology and goes too far in this, sometimes denying any reality to substance, sometimes giving an impression that movement can disengage from substance and behave as if it were some kind of independent reality. The view indicated by Lacey, where we have substance in motion, seems more plausible. The moving and acting party here comprises substance and motion together, and the division of matter into real motion and the illusory substance does not work.

To stress that the movement involved in matter and consciousness is different, we must also look again at the distinction between irreversible and repeatable change which Bergson made in the beginning of *Creative Evolution*[12] and our observation that not all temporal relations have qualitative significance.[13] Repeatable changes, and events which do not affect each other, are not part of heterogeneous duration, where all elements are entwined. This mutual independence belongs to entities whose inner nature is not essentially tied to particular temporal coordinates and particular links with other entities, and this can be said to be a feature of the material world. We observe duration in the inanimate world as concrete instances of movement from one place to another, and as internal movements within entities, but material entities and processes can be isolated from each other in space and in time and may have no effect on each other's nature. Thus inanimate reality does not illustrate nor does it imply universal duration. Bergson presents movement in matter as a case of duration but it lacks the most essential feature of duration, namely growth as an inner accumulation of its own history, which is found in the higher being of life and consciousness. This makes it difficult to talk of material movement as duration at all, and reveals a significant discontinuity in the Bergsonian picture.

The movement of displacement obviously serves as a prototype for consciousness in Bergson, and the relation of substance and motion as a

12   See Chapter 4, Section 1.
13   See Chapter 6, Section 1.

prototype for the relation of brain matter and consciousness. However, it cannot be any more than a prototype: the movement caused by two objects pushing each other cannot explain the supposed movement from consciousness to matter and from matter to consciousness as if they were equal poles. Taken in itself, matter cannot account for links with life and consciousness.

3   Living Matter

When matter is taken as a component of living structures, its role changes. This change does not come from within matter per se but from the more complex biological heterogeneity of which matter is only a component. In the biological structure matter gains new functions which can only be detected at this level, where the more complex structure includes and accounts for a simpler structure within itself.

Describing the biological realm, Bergson talks of the creative force of life which launches itself into matter and 'strives to introduce into it the largest possible amount of indetermination and liberty' (*CE*, p. 265). The relationship between the impetus and matter is not straightforward. On the one hand, the impetus needs matter for its own realization and on the other hand, matter restricts and channels its expansion. Matter becomes indispensable material for life and an agent restricting the expansion of life at the same time – a typically Hegelian contradiction, with a thesis (matter is material for life) and an antithesis (matter restricts life). In the living organism the two opposite movements of matter and life meet and are reconciled, interact and together result in an 'organized' body.

Every manifestation of life is a creation of something new. The novelty is brought in here by the impetus itself, as it effects *creation* as against the mere rearrangements of parts. We can observe that matter ensures that the body *exists* as a physical object, and the impetus is responsible for it being *created*, for its development in time, for the novelty of forms.

In Bergson's own theory of the two opposite processes, life is 'an effort to remount the incline that matter descends', and if in life he finds a creative power, in materiality he finds features of 'a thing *unmaking itself* (*CE*, p. 258). The creative process of organic life includes a struggle between the creative force and matter which serves as material for the constitution of living bodies. Creative force prevails when the body lives, but as time passes, the material base becomes damaged, gradually or suddenly, and life is unable to continue in a body that refuses to live. Just as creative force cannot apply itself to a fragment of inorganic matter, i.e. animate a stone, it is equally unable to thrive in a body that ceases to be organic, whose internal structure tends towards passive inert existence.

Bergson points out that life in general is little concerned with the indefinite preservation of an individual organism. Perhaps this just demonstrates the limitations of the living tendency and its dependence on suitable material for its own realization. The living tendency is strong enough to create a body, but not strong enough to make it grow indefinitely in size or let it persist indefinitely in time. Eventually, matter in its passivity prevails, but this is only a localized victory because life is carried forward by means of other bodies. This is what Bergson must mean when he says that life could resist 'the most formidable obstacles, perhaps even death' (*CE*, p. 286). The lack of interest, or the lack of obsession at least, that life demonstrates towards individuals, makes an individual death insignificant in regard to the whole picture, because the individual organisms are merely fragments of one living world.

We see that an organism dies because disintegrating matter can no longer serve as the physical base for life, and not because the living force at some point expires. Bergson's observation that an individual life is so easily terminated as against life in general which seems indestructible, can be explained by the fact that an individual life, cut off from the general tendency, is weaker and more vulnerable than the general tendency and, dependent on the concrete fragment of matter, lives in this hypostasis for as long as matter is able to support it.

Bergson stresses that the living tendency is a tendency to individuation, but he does not unpack the role that both matter and the creative push play in this process. He implies, though, that it is the creative tendency that is responsible for all positive modifications of the resulting body or organ. But, as we can see, the living tendency in itself is indifferent to any individuation. It uses matter to promote itself further and further. But that fragment of it which is actually tied with matter is affected by its concreteness, for matter is always concrete, so the living tendency becomes concrete too. Affected by matter, it no longer participates in the whole of the living world to the extent of disregarding the interests of any particular individual. Instead, having constructed an individual and being part of it, concretized and individualized, the living tendency is cut off from the rest of the living world so that when the matter that it has organized becomes destroyed, this individualized life is destroyed with it, and the persistence of life in the world is not going to help that particular life which depends on the integrity of its concrete materiality. It seems that, contrary to Bergson, it is matter that gives the living tendency foundation for individuation, as it brings in concreteness at a physical level.

The closer we look at organic matter, the more we realize that the role of matter in this compound is more important than Bergson wants us to believe. We realize that the impetus of life is solely responsible for the push, for the advancement and growth in a very general sense. On its own, the living force is impersonal, indifferent to any concreteness, lacking any particular direction, and responsible for no concrete features in the living organism. Planted in matter, it pushes its organic growth, but the rest – the individuation of the living organism, its generic and unique features, its welfare and its death – all this is determined by matter which exists as a concrete carrier of the tendencies that it shapes, as a repository of life, as its protector and as the immediate producer of death. Without matter, the living tendency is a pure possibility, it is nowhere and everywhere. When realized in concrete bodies, it becomes embedded in them, localized, and acquires co-ordinates in the physical world. That part of it, which remains a tendency, also remains nowhere and everywhere, until it concretizes itself in future forms. It seems that the creative impetus is a general principle responsible for all possible manifestations of life, from the creation of a cell to the pattern of behaviour of a concrete animal. Any specifications observed in tendencies where the impetus gains particularity must be due to the nature of matter that the impetus encounters and which shapes its performance. Without being involved in a concrete manifestation, the impetus is indescribable and unimaginable. But, we shall say again, this new significant role of matter cannot be identified as a potentiality when matter is considered prior to its involvement in biological existence.

4   Psychological Duration

The heterogeneity with a highest complexity (consciousness) accounts for simpler, biological and material levels in perception. To illustrate this phenomenologically, we will look at the feeling of warmth – one of the very first components of psychological duration ever mentioned by Bergson (*TFW*, p. 1).

When we place our hand in hot water, from the standpoint of the material world there is physical contact between two material elements – the flow of water and a soft, flat oblong object with five protruding extensions. As members of the material world, neither the oblong nor the water is competing for preferential treatment. The situation is decentralized with no focal element as to for whose sake the physical contact could be taking place. After a prolonged contact with the water, the temperature of the oblong rises. If the temperature of the water becomes forty-five degrees Celsius, the oblong turns red, but this does not change the axiologically neutral status of the situation, for no axiological statement has any reason to appear in the material world.

From the standpoint of the living world, the oblong is a palm, and the extensions are thumb and fingers. It is now part of an organism, an organ of holding, selective touching, exploring and manipulating matter. Water, too, changes its meaning. It stops being merely a colourless transparent liquid compound of oxygen and hydrogen and acquires a new role – as part of the environment. The situation is no longer decentralized: it acquires an axiological element which determines its centre, a step towards subjectivity, we may suggest. Neither an oblong nor a colourless liquid can experience pleasure or pain, but the hand can, and water as part of its environment can be a cause of pleasure or pain. Also, the hand is an organ designed to perform certain functions. These functions which constitute the hand-ness of the hand can be affected by feelings of pleasure and pain resulting from physical disturbance: if the temperature of the water increases to 45°C, the water becomes a hostile environment, as it would scald the hand, damaging its hand-ness. As a feeling entity, the hand becomes the centre to which the criteria of good and bad can be applied. Water, not being another living entity rivalling the hand's right of centrality, becomes a means to promote the hand's existence or a potential threat to that existence.

The mechanical contact of the water and the oblong which turns out to be a dialogue where the living hand draws either support or destruction from the mineral, becomes at the level of consciousness something else. The value that the hand assigns to itself is shifted and reassigned to its conscious bearer – the human being – with the hand losing its value as the ultimate end but retaining its value as a functional part of the human body. In cases when the hand forfeits its hand-ness, for example when it is dangerously infected, it may be amputated if the survival of the whole organism demands this.

As a functioning part of the human being, the hand's pleasure or pain becomes pleasure or pain for the person. The sensations that I am aware of are localized and restricted to one single area of my body and I am fully conscious of the restricted location of the sensation, but the pleasurable feeling spreads to my body as a whole, and delivers pleasure to me as the self. The pleasure is not identical with the feeling of warmth and yet it is not separated from it. I am certain that although I can localize the sensation and maybe even cannot help but localize and hence limit it, I cannot separate it from myself, from my 'I' either: the sensation is experienced by me directly as the warmth of my hand.

The hand as oblong or as organ is not part of my psychological duration. Yet my mind experiences directly the sensations of my hand; not learning about them theoretically, but gaining factual knowledge of warmth. My mind understands intimately what the hand is in mechanical terms (its shape, size, weight) and what it is in biological terms (a functional part of my body). My knowledge of these aspects may be inaccurate and incomplete, but my mind does not question or doubt the data it receives; it does not question whether the water is hot or cold, wet or dry – it definitely *knows* one way or another.

It seems that my pre-reflective mind has direct access both to the material and biological nature of my body, and that my mind causes the awakening of both inert and living matter to another mode of existence – as being experienced, being known. My mind, which is not an extended entity, experiences extension, location, direction, size, weight, colour, and density. My mind, which is not an organ, experiences warmth, pain and pleasure. Psychological duration appears to be a higher process which includes all lower processes, not as empirically given, but as empirically known. Psychological duration in its pre-reflective state comprises dematerialized but otherwise real material and organic processes – as images, we should perhaps say.

Referring back to the Bergsonian theory of images, this is perhaps the best understanding of 'images' we can have – as dematerialized events, and distinguishing them from material events themselves. Then perception, we must admit, dematerializes embodied reality, and where Bergson sees the node of continuity between the perceiving mind and the perceived matter we nevertheless see a discontinuity – a jump from the empirically given to the empirically known.[14]

5   Heterogeneous Self

The involvement of matter and organic being in the feeling of warmth contributes to the content of our psychological duration. We have observed them in a certain order: material characteristics came first, biological ones followed and psychological ones concluded the process by including them all. But in the light of the conclusions made in Chapter 5, we will emphasize

---

14   For an exposition of Bergson's theory of images see Chapter 3, Section 2.

that, as a fragment of heterogeneous duration, our sensation of warmth cannot be separated into a sequence of sensations whereby we first feel the shape of our hand, then its rising temperature, then the pleasure. The awareness of all these characteristics appears jointly as one experience, and we separate various sensations by the different effects they produce on our body and mind. Here we see heterogeneous duration at work: characteristics as diverse as shape, temperature and pleasure are united in a single awareness of warmth; they can be singled out as qualities which emanate different relations but cannot be considered in a sequence as separate elements. Thus a single sensation illustrates a unity, a continuity of mind and body, whereby bodily features are included in the mind in the disembodied form as tactile images.

The idea of heterogeneous duration comprising diverse qualities in a temporal flow is able to accommodate a potentially unlimited number of diverse phenomena. Within the theory of self as duration, we do not search for the core of selfhood eliminating the rest of the human being from the picture, but talk of the entirety of human existence, as various aspects of the human being do not exist separately from each other.

We must stress that the 'highest' psychological duration including other 'lower' durations involves discontinuities when we consider this connection from the point of view of the lower durations: inanimate matter suddenly becomes a base for life, and life suddenly gives rise to consciousness. But lower durations taken prior to their involvement in the entire heterogeneity of the human being cannot account for this entirety and for other elements that are involved in this heterogeneity – just like the isolated note *d* cannot account for the musical sequence *abcda*.

We recognize that the direction in which we are following Bergson is different from the traditional Cartesian position taken up by Hume and Kant, philosophers for whom the self, the 'I', necessarily meant the thinking subject, the author of thoughts, as opposed to the object of those thoughts. Bergson dissolved the opposition of subject and object, and so made the problem of the self shift from the concern about the rationality displayed by the self, to the being of the self which includes rationality as its part: the idea of the self that *is* precedes the idea of the self that *knows*. We still acknowledge the privileged role of waking consciousness as a summit, so to speak, of the human being, but we cannot pretend that it has some independent existence and assert that the self simply equals a knowing mind. Adapting Loizou's distinction of 'thin' and 'thick' conceptions of the self to our discourse,[15] we can say that whereas the Kantian and Humean are amongst the 'thinnest', where the self is equated exclusively with the rational knowing subject, the Bergsonian supersedes the thickest possible conception of the self: according to the Bergsonian principles, the human being is irreducible to any of its components, and the self comprises psychological life and the body, taken in their past and present existence.

Concerned with the ultimate principle of being, Bergson did not offer a systematic view of selfhood as the self is only a fragment of being, but there is enough evidence in Bergsonism to construct a theory of heterogeneous self. As duration impregnated with memory, the self comprises all of its history, both as a psychological and as a physical duration. Human beings, developed from inanimate matter to bearers of intelligence, encompass everything they have accumulated along the way: physical laws that govern inanimate matter, instinctive growth of the living tissue – these layers underline the existence of our mind. Everything that affects inert matter will also affect living matter which is built on inert matter, and features of living matter will be included in the existence of a conscious being.

The self as duration is saturated with its own past, and its every manifestation, from the existence of an individual to an act performed by the self, is the work of all of its history, going back far beyond the birth of the individual into the history of the living world and of the universe. It could even be said that the human self is a culmination of all worldly processes – from movement of the substance in inert matter to the advancement of life in organic bodies.

Part of the Universe and a fragment of evolution, human consciousness or psychological *duration* is the continuity of interrelated psychological events. The mechanism that binds them together is *memory*. *Perception* supplies new material for the content of duration and is given meaning by means of the information about previous experiences preserved in memory,

---

15   Loizou, *Time, Embodiment and the Self*, 70.

revived by *recognition* and grafted onto the newly supplied sensory data. This complex process of cognition, involving the joint work of perception and memory in recognition, is a base for the work of *intelligence* which, for Bergson, is ultimately a manufacturing ability, i.e. an ability to manipulate matter at will, decompose and recompose it in a multitude of ways. Apart from the explicitly rational and logically narrative side of our conscious life, represented by intelligence, there are also feelings, emotions, beliefs and acts of volition.[16]

Instinct too comes into the life of our self. It is a prolongation of the biological drive which is involved in the construction of organs and in their functioning. It is evident in our involuntary activity (e.g. breathing) but also included in our rationalized actions (e.g. procreative behaviour). In addition to intelligence and instinct, there is also *intuition*. If intelligence is the dominant faculty of our psyche and the role of instinct in humans is restricted (unlike in animals which survive mainly due to the work of instinct), intuition as a cognitive faculty exists in a nascent state and manifests itself really only through artistic work. Like instinct, it is involved in direct creation, where the complex structures are produced in one act of inspiration without the rationalized analysis and composition of their parts. But its goals go far beyond the goals of instinct, which is only concerned with the immediate tasks of the preservation or promotion of organic life. As a cognitive and creative faculty, intuition is disengaged from the practicality of intellectual knowledge or manufacturing and is able to grasp the very being of things.

---

16   Independently of 'thick' and 'thin' conceptions of the self, there are theories that define self as a narrative structure, involving temporally extended eventuality, and those that prefer to see it as episodic, confined to the present moment. (For an example of the former see Crowther, 'The Cohesion of the Self' in his *Philosophy after Postmodernism: Civilized Values and the Scope of Knowledge*, 78–100, and for an example of the latter, Galen Strawson, 'The Self', in Shaun Gallagher, Jonathan Shear, Galen Strawson, eds, *Models of the Self* (Thorverton UK: Imprint Academic, 2000), 1–24.) The Bergsonian concept of the self is, of course, comparable to the self as narrative, but it supersedes it, because in Bergsonism, we also take into account the past history beyond the concrete existence of the self such as evolutionary development.

The mind, comprising various tendencies and processes, is embedded in the processes and tendencies of its physical base – the organic body. The body is duration of organic events which constitute the life of the organism. They are bound together by memory, just like psychic events, but in a different way: memory of living matter does nor preserve images of events but preserves extraction of concrete events as tendencies, which are manifested later in subsequent events. Duration of a living organism also involves behaviour of that organism, which comprises movements as a result of choice made by the organism. Also, the life of the organism in itself is part of the mega-process of evolution, and as such contains tendencies remaining from the biological events of all preceding ancestral life forms. It could be said that the relationship between organism and its past is direct rather than representational.

Living matter is matter that supersedes inanimate matter, but includes the latter within its own structure and thus the duration of inanimate matter is also included in the structure of the animated matter. Just like inert matter, organic matter, on a sub-cellular level, involves internal vibrations and vortices. In addition, we must not forget that the motion of a living organism is not exhausted by its behaviour; bodies, just like inert objects, can be pushed, lifted or thrown, or in other words, involved in mechanical motion which differs from deliberate actions.

The human being, as an heir to all its preceding history, is a multidimensional being comprising features of inert matter as well as those of organic matter. For example, a human being possesses qualities such as size and weight which he or she shares with all objects extended in space. 'I' as a physical object is a dimension of the self different from 'I' as an organic body and from 'I' as a sentient being. We accept the totality of the self as a material, biological and psychological unity, bearing in mind that the self can function in all three dimensions: as a material, biological and psychological entity.

Amongst material objects, the human being is but another material object, and those sides of us which do not find counterparts in material things, are left out of the scene. We are inert matter for material objects, and our interactions within their realm are restricted to mechanical displacements: pushing, lifting, dropping, breaking.

In the living world, our biological side finds response. Living existence involves material existence, so we can still push, lift, drop and break, but we also engage in biological interactions: we eat, kill, grow, procreate. Our biological actions retain their material significance, however, so that the events of the organic world acquire a twofold meaning: the same event, in the material dimension would be pushing, whereas in the organic dimension it could be causing pain, and increasing in size, for a material object is growing if this object is an organism.

In the psychological dimension of the self, the latter displays features, specific to consciousness, but also retains both biological and material ones. For a phenomenological example, we can recall that alongside the feeling of warmth, Bergson mentioned the feeling of sadness (*TFW*, p. 1). Whereas analysing the feeling of warmth we found that it consists of consciously acknowledged material and biological components, the feeling of sadness contains an extra layer which is generated in consciousness and not included in the previous, biological, level.

The feeling of sadness can be triggered by, say, a letter. Then, just as in the case of warmth, we have a purely physical element to begin with: the flat square piece of paper in the proximity of a sphere which turns out to be the head, because being a head involves being a sphere. On the biological level, the eyes perceive the paper from the distance of some thirty cm and distinguish lines of black marks. The further development of sadness involves memory. Our mind recognizes black marks as letters of a familiar language, groups of letters as words. Then we engraft further layers of meaning, such as emotional connotations, onto the words of the letter. We connect the concepts in the letter with our memories, and their combination evokes the sadness.

Despite its physical and biological manifestations, such as a lump in our throat and tears in our eyes, we understand that sadness, unlike the feeling of warmth, is necessarily generated by consciousness. In the feeling of warmth, the role of the mind is limited to the acknowledgment of the biological event. We mentioned the feeling of pleasure associated with it which is more like a feature of consciousness, but this still could be a simple acknowledgment of what is beneficial for our body. As for sadness, even though we find both physical and biological elements involved in the causing and manifesting of this feeling, the actual production of it is obviously the work of the mind, or soul, or psychological duration. When we say 'I am warm', we recognize that it is the body that is warm, and that the psychological being warm is an acknowledgment of the bodily experience. When we say 'I am sad', there is nothing apart from our psychological self that can be sad, because our body cannot be sad. Sadness, of course, can be depicted by an actor via a bodily performance, and real sadness projects itself in our appearance, but cannot be reduced to it and originates in the psychological realm.

Of course, duration, although shown in *Time and Free Will* as purely inextensive and spiritual, could not exist as such in reality because its existence presupposes growth, the addition of new content, and the material for new content comes from the physical world. In *Matter and Memory* Bergson shows that data about the world is delivered to our consciousness by our senses in such a way that we experience the very materiality of the world – we partake of it as recipients of the direct influence that weight, size, colour or sound produce on our body. In this sense, psychological duration partakes of the materiality and extensity of the physical world since its content comprises sensations that originate from a physical cause. 'It is in very truth within matter that pure perception places us, and it is really into spirit that we penetrate by means of memory. But, on the other hand, while introspection reveals to us the distinction between matter and spirit, it also bears witness to their union' (*MM*, p. 180).

In *Matter and Memory* Bergson nevertheless talks about 'reconciliation between the unextended and the extended' (*MM*, p. 181). This reconciliation (which appears to be inconsistent with the theory of pure duration in *Time and Free Will*), takes place in pure perception:

> We place the perceived images of things outside the image of our body, and thus replace perception within the things themselves. But then, our perception being a part of things, things participate in the nature of our perception. Material extensity is not, cannot any longer be, that composite extensity which is considered in geometry; it indeed resembles rather the undivided extension of our own representation. That is to say, the analysis of pure perception allows us to foreshadow in the idea of *extension* the possible approach to each other of the extended and the unextended. (*MM*, p. 182)

We observe that our mind is aware of the physical qualities of things in such a way that it does not just know *of* them – our mind has a direct experiential knowledge of physical things. Unextended, it experiences their extensity in sensations delivered to it by the nervous system of the body. Thus extensity and inextensity must partake of each other: inextensity as a recipient of the physical influence and extensity, as a bearer or producer of a visual, audio or tactile 'image'. This is how one can understand Bergson's position regarding the unity of spirit and matter occurring in perception. Memory helps us in this, he believes: 'Our conception of pure memory should lead us, by a parallel road, to attenuate the second opposition, that of quality and quantity' (*MM*, p. 182).

Bergson demonstrates that what spirit and matter have in common, what makes them belong to the same order of reality and what makes their interactions possible, is that they both share in movement, to the extent that both are, primarily, movement themselves. In perception the extended and the inextended flow into one another. 'That which is given, that which is real, is something intermediate between divided extension and pure inexten-sion. It is what we have termed the *extensive*' (*MM*, p. 245). The extensive participates both in things (when they are taken out of an abstract space) and in our consciousness as we perceive extensity.

However, although in *Matter and Memory* Bergson accounts for the continuity of subject and object in his theory of perception, the perception of an object does not explain the emergence of an emotion.

Here it may be useful to recall Bergson's 'deep-seated phenomena, the cause of which is within us and not outside' (*TFW*, p. 5). This could imply a division of psychic phenomena within psychological duration, where the feelings of sadness and warmth could be labelled as more or less deep-seated. Sadness, even when triggered by outer circumstances, is not simply taking the physical and the biological phenomena to a higher state of their being known. Sadness appears as a result of the reader coming into physical contact with the letter, but it is not a mere prolongation of this physical contact: as sadness, it is a new creation within the psychological duration per se, which reveals another split in duration.

CHAPTER 8

The Structure of Rational Selfhood

1 The Pre-Reflective Self

We do not expect rational subjects to be all the same, as every person demonstrates differences in temperament, values and preferences. Bergson recognizes that one's rationality is not defined solely by self-conscious awareness but includes character as well – the character which is built over time: 'What are we, in fact, what is our *character*, if not the condensation of the history that we have lived from our birth – nay, even before our birth, since we bring with us prenatal dispositions' (*CE*, p. 5)?

For the analysis of the pre-reflective base of psychological duration we can use terminology introduced by Loizou who presents psychic dispositions, values, traits of character, preferences as constitutive attachments – the self's inner objects which may not be present to one's consciousness but are not detached from the self as, say, objects of sight: they constitute our character and our pre-reflective self which is a base for rational performance.[1] At the pre-reflective level, 'constitutive attachments define not merely what I will, what I do, or what I have, but what I *am*'[2] and give our subjectivity colour and body.

Constitutive attachments do not enter our psychological duration at a definite time as concrete temporal episodes. They are present in our psyche, and there must have been time when they were not present, but we cannot

---

1   Loizou, *Time, Embodiment and the Self*, 67–70. Loizou introduces 'inner object' only as a provisional term, but it effectively highlights the distinctive and problematic nature of character traits.

2   Loizou, *Time, Embodiment and the Self*, 70.

account for a definite date when such and such preference or belief came to exist and to contribute to one's psychological duration.

Tendencies that form our character[3] are not evident unless they are manifested through some concrete action. One's love for music is always there but is not manifested unless there is, say, a choice between going to a football match or to a concert – and the lover of music will choose the concert, thus revealing the preference for musical entertainment as a trait of character.

Creating a tendency must be akin to acquiring a motor memory.[4] A tendency exists prior to its realization but, before it can exist as unrealized tendency, first it needs to be formed. To understand the mechanism of forming a tendency, we shall recall Bergson's assertion that '*the formation of memory is never posterior to the formation of perception; it is contemporaneous with it*' (*ME*, p. 157). It is an illusion, Bersgon says, to believe 'that memory *succeeds* perception' (*ME*, p. 160). Memory, he asserts, is formed alongside perception (*ME*, pp. 159–60), and after perception has ceased, memory remains (*ME*, p. 164). 'The memory seems to be to the perception what the image reflected in the mirror is to the object in front of it. The object can be touched as well as seen; acts on us as well as we on it; is pregnant with possible actions; it is *actual*. The image is *virtual*, and though it resembles the object, it is incapable of doing what the object does' (*ME*, p. 165).

Image remembered is then an object dematerialized – as concluded in Chapter 7.[5] Memories that we keep and which constitute our psychological duration consist of dematerialized eventuality. While we are engaged in an event which contributes to our acquiring a tendency, our physical and biological interactions are accompanied by their dematerialized counterparts. When the physical and the biological cease, the dematerialized remains. And, according to Bergson, it remains even though we are not

---

3   For Bergson, character is a network of tendencies ('that special tendency that we call our character' (*CE*, 104)).

4   For an exposition of motor memory see Chapter 3, Section 3.

5   Chapter 7, Section 4.

conscious of it: this memory is always there and always ready to push its way into the consciousness.

A tendency thus consists in an inactive memory of numerous instances of repeated events of the same type. This memory is not always evident but, as an indestructible part of psychological duration, the tendency contributes to our character and to our self at all times as unconscious, unremembered memories. Habit memories are thus traces of past events, where events have become anonymous with their dates and places erased, being irrelevant to the formation and realization of the habit.

We do not remember each instance of eating and enjoying spicy food, for instance, but these inactive memories condense themselves into a tendency to like spicy food, with this tendency becoming part of our character. Enjoyment of eating spicy food on particular occasions becomes the reason for choosing it in consequent situations. When the circumstances permit this preference to be revealed in the act of making a choice of food, this memory, de-concretized and transferred into a generalized tendency, becomes manifested. Its pastness is being restored: we can admit that our reaching out for spices is the evidence of our having acquired the taste for it in the past.

But what about the ontological status of inactive memories when they are dormant? They are not extended, not materialized, not obvious, not in any place, but they exist, according to Bergson, and constitute the past. Bergson calls memory in a state of inactivity pure past, but it may even be appropriate to regard it as a-temporal, as it loses any connections, any relations with reality: it is there, but without a date in time, or location in space. The property of being past of such memories is only revealed when they are remembered or brought to light as actualized tendencies. When they are not remembered they can only be understood in negative terms. Since unremembered and unrealized memories are inactive, they cannot interact with each other, because interacting would be acting. The status of inactive memories is dubious: they are part of duration (according to Bergson) but, having lost all their concrete characteristics, they have lost everything that ties them to the reality of concrete duration. Bergson does not seem to offer an explanation of this difficulty.[6]

Another difficulty consists in Bergson insisting that all our past is preserved and that the demands of the current perception retrieve those memories that are suitable and relevant.[7] But there is no criterion of their suitability and relevance, so in principle memories can be retrieved at random. Secondly, if the entire past contributes to the formation of every present, then we have an undifferentiated totality. Contrary to Bergson, there must be a division of experiences into influential which contribute to my character and non-influential which do not, even though the principle of selection is not immediately obvious.

2   Unity of the Self in Self-Consciousness

Moving on to the questions of identity, of the self acknowledged and rationalized, we will note that if Descartes, Hume and Kant were beginning their discourse of the self with the 'I think' we, following Bergson, arrive at the 'I think' only after considering its material, biological and unconscious foundations.

How does the self become self-conscious? When dealing with the self's rationality we enter the domain of the obvious, of the phenomena. However, we are not interested in the acting self yet but in that fragment of the process of selfhood that lies between the pre-reflective self and the self's physical engagement with the external world. This is rational self-consciousness, alert and waking phenomenal mind comprising thoughts, desires and ideas, the

---

6   As I demonstrated in Chapter 6, the vaguer the spatial parameters of an entity are, the vaguer are its temporal coordinates and its ontological concreteness too (see Chapter 6, Section 2).

7   For my division of the past that survives and the past that 'drops off' see Chapter 6, Section 5.

'I' of the individual. It is there that we will look for identity – conscious, acknowledged self-sameness preserved over one's entire life.

How would Bergson explain the unity of temporally stretched self? By the work of memory, of course: all our experiences are preserved, explicitly remembered or not, so that we comprise in our now-existence all that we have ever experienced in the past. We have noted above our suspicions regarding the significance of the entire past for the person, but granted, this accumulation of experience, de-materialized but not lost, explains the temporal unity of the self. But does it explain this unity *to* the self? Most of the preserved memories are suspended in a non-remembered state, and can be traced back to the point of their origin only retrospectively, when they are brought to light in an act of remembering. Thus the self is not aware of its largely unconscious entirety, and yet it is aware of and believes in its own self-sameness preserved over the time.

The selfhood analysed in section 1 of this chapter is not yet self-consciousness, as self-consciousness involves reflection of consciousness upon itself, and this is what the pre-reflective self is lacking: one can be kind, angry or love music without acknowledging one's self-sameness. Turning our attention to self-consciousness, we will now consider it in terms of rational awareness – an ability to navigate between various temporal planes and synthesize temporally diverse events into a unity of one's own 'I'.

Should we regard the 'I' as merely a concept or as another, possibly central and privileged, process of selfhood? As a concept, we would expect it to be an abstraction of all processes of selfhood, to be superimposed on the diversity of personal events as an a-temporal entity which represents and points at one's identity, but which has no correspondence to anything real.

It is this a-temporality of the conceptual 'I' that Hegel uses in his discussion of absolute religion where he sees the possibility of entering heaven, not after death, but during one's life, provided one loves God and detaches oneself from empirical reality.[8] This would mean taking the con-

---

8   Hegel talks of consciousness of reconciliation, abstracted from present actuality
     (G. W. F. Hegel, *Lectures on the Philosophy of Religion* (Berkeley, Los Angeles, London:
     University of California Press, 1988), 460).

cept of 'I' as primary reality and identifying oneself with this concept. Thus, dematerializing oneself to the extreme, one can hope to experience immortality as immateriality whilst alive, for the concept will remain indefinitely. Bergson would argue against this view of immortality, for concepts cannot be immortal simply because they have never been alive: he asserts the concrete reality of processes of selfhood. So, if we recognize the 'I' as a concept, we need to find out what this concept means to us when we use it: does it stand for a specific process of selfhood or could it be, in ontological terms, made redundant?

Bergson's explicit reference to the self as 'I' can be found in *Mind-Energy*:

> Besides the body which is confined to the present moment in time and limited to the place it occupies in space, which behaves automatically and reacts mechanically to external influences, we apprehend something which is much more extended than the body in space and which endures through time, something which requires from, or imposes on, the body movements no longer automatic and foreseen, but unforeseeable and free. This thing, which overflows the body on all sides and which creates acts by new-creating itself, is the 'I', the 'soul', the 'mind', – mind being precisely a force which can draw from itself more than it contains, yields more than it receives, give more than it has. (*ME*, p. 39)

Yet Bergson does not dwell on the issues of self-consciousness. It seems that, for him, there is no urgency in distinguishing between consciousness and self-consciousness: both are understood as consciousness, as Bergson concentrates on the function that conscious processes play in action and manufacturing. From this point of view, self-consciousness is consciousness that is just more highly developed and accommodates a wider scope of possible action. However, Bergsonism, does not deny self-consciousness – Bergson's position on it can be retrieved out of his deliberations about psychic states, memory, intelligence and intuition.

3 Bergson: Intuition versus Analysis

Self-consciousness is self-knowledge, an awareness of one's 'I', where the most basic statement of selfhood, the 'I am', is acknowledged and taken for granted. Like everything else in the Bergsonian system, self-consciousness must be understood as a process with irreducible temporality.

Self-consciousness, awareness of one's own unity as a temporal individual, is an achievement and a case of rational thinking. Therefore in order to answer the question as to whether we could trust our rationality to deliver a reliable portrayal of our self, we ought to evaluate rationality in general. In Bergson we find two ways in which rational awareness manifests itself: intelligent analysis (a practical means that helps manipulate matter) and intuition (a direct grasp of reality).

The idea of intuition is developed and contrasted with analysis in Bergson's *An Introduction to Metaphysics*, but to understand fully what this contrast entails, we need refer to Bergson's subsequent work, *Creative Evolution*, which describes intuition as 'instinct that has become disinterested, self-conscious, capable of reflecting upon its object and of enlarging it indefinitely' (*CE*, p. 186).[9]

From the analysis of *Creative Evolution* we remember that instinct, intelligence and vegetative torpor are independent tendencies of the original impetus of life which is carried on throughout the evolutionary development from the most primitive forms of life to plants, to Hymenoptera and to man, where vegetative torpor, instinct and intelligence appear in their most developed forms. Whereas intelligence is the ability to dissect reality in order to, ultimately, construct man-made tools, maintaining man's work of adaptation to the environment, instinct is nature's ability to create natural tools, organs and organisms, in a simple act of creation. This work of creation is prolonged then into the instinctive use of an organ, and into the instinctive behaviour of the living organism which acts as if it knows the outcome of its actions. From this perspective, instinct is creative power

9   For an exposition of Bergson's theory of cognition see Chapter 4, Sections 3–6.

combined with direct awareness of the happening, so that the life of any individual organism shows traits of conscious behaviour. Adaptation to the environment – living 'at the expense of other animals', or 'a tendency of soft organisms to defend themselves against one another by making themselves, as far as possible, undevourable' (*CE*, p. 137) – resembles a conscious response to external circumstances,[10] and is an indication that living bodies are aware of possible dangers or of opportunities to use other bodies for food.

This 'knowledge', this quasi-conscious awareness observed in the animal world and beneficial for the whole of the species as it develops, cannot be passed from one individual to another via a signalling system such as human language. This knowledge that is involved in the living organisms and that has a direct effect on the evolutionary process, must be of a different type than that which we have as conscious beings and must have a close connection with the ontological structure of the being. It must be part of the ontological structure of a living being, if having this knowledge *in itself* allows the living being to maintain or alter the form of its descendents. So, the tendency of life that Bersgon talks about, must be a force embedded in each living organism that is knowledge on the one side and the power to act on matter on the other, with knowledge directing physical power.

It is here, more than at any other point in Bergson's philosophy, that he advocates existence merged with epistemology where information about the being constitutes part of the being, and the process of being includes an element of knowledge, so that the knowledge is passed on through being,

---

10   This resemblance is based on the apparent similarity of their functions: both consciousness and instinct provide successful responses to circumstances. But following William James, one can say that although an animal acts as if it subserves abstractly understood purposes such as self-preservation or defence, its actions lack any understanding and the animal acts simply because it cannot help but behave in this way. (See William James, chapter 24 'Instinct', *Principles of Psychology* (New York: Courier Dover, Vol. 2, 1950), 383–441.) If successes of instinctive actions imply knowledge, this knowledge is not explicitly given to the acting agency itself because the animal does not connect its own effort with the success. This 'knowledge' does not have a knowing subject. That is why Bergson talks of 'resemblance' and of 'quasi-consciousness', avoiding a direct equation of consciousness and instinct.

itself as part of the process of being. This correlates with Bergson's claim that ontology must include psychology, and that the true task of metaphysics consists in remounting the incline that physics descends, bringing back matter to its origins, and building up 'progressively a cosmology which would be, so to speak, a reversed psychology' (*CE*, p. 219).

This is where intuition originates – in the instinctive knowledge embedded in biological being. Bergson indicates how instinct can develop into intuition: 'Instinct is sympathy. If this sympathy could extend its object and also reflect upon itself, it would give us the key to vital operations' (*CE*, p. 186).

Intuition, i.e. instinct taken to a higher level, is a kind of awareness where we, supposedly, immerse ourselves in the object and discover its nature directly. Even though we have entered the object from outside, we forget the moment of entry and we experience the nature of the object as if it were part of our being, like pain or thirst. As in instinct, our knowledge then is a reverse side of being, but what makes intuition different from instinct, it seems, is that, whereas in instinct one's knowledge is bound with one's being in order to preserve and promote it, in intuition our knowledge becomes disengaged from our being and is bound with the being of another entity, and this is why intuition is described as disinterested instinct: it no longer looks after the biological structure of the organism but provides organic knowledge of another thing or being without it being linked directly to the process of biological creation. This epistemological faculty, according to Bergson, has not been fully developed in man; we only get glimpses of it in art where the artist's genius creates a complex and varied masterpiece in a simple act of inspiration, and also in the act of aesthetic contemplation.[11]

---

11   Simondon, developing the idea of intuition further, suggests placing creative intuition in a wider context rather than restricting it to aesthetics, and claims that it is present in manufacturing as well – in the form of invention (Pascal Chabot, *La Philosophy de Simondon* (Paris: Vrin, 2003)).

To summarize: both in intuition and in instinct one's attention is directed at the essence of the object[12] to the extent that there is a merging of attention of both the subject and the essence of the object, but in instinct, one is already within the object, and in intuition, one needs to place oneself within the object. Instinct forms part of the creative process but intuition itself does not create, and this is linked to the fact that instinct is bound with one's existence whereas intuition is disengaged from one's existence and bound with the existence of the other being or thing. Also, instinct acts involuntarily as it is imbedded in the organic structure of the organism, but intuition acts only when it is willed, as it is a result of an effort. Even though Bergson limits examples of the use of intuition to artistic creation, he indicates that intuition could in principle give genuine knowledge of any object, which would also include knowledge about our self in self-consciousness.[13]

Unlike intuition, analysis is the conceptual approach to knowledge, in which we dissect reality, looking at it from the outside, and rearrange it according to our conceptual framework. If in intuition we enter into the object, in analysis we move round it. Intuitive data does not depend on a point of view nor rely on any symbol, but data received via intellectual analysis depend on a point of view and rely on symbols. Intuition gives an absolute knowledge, grasping what is unique and essential, whilst analysis provides relative knowledge expressed in concepts. In intuition, we coincide with the object by means of intellectual sympathy, and in analysis we reduce the object to familiar elements and then attempt to reconstruct the original. Intuition thus is a simple act and analysis can go on indefinitely. Bergson finds analysis justifiable in natural sciences, but he would prefer metaphysics to use intuition, because intuition deals with the real, moving thing, whereas analysis crystallizes and immobilizes the moving object.

---

12   It may be argued that in instinct, attention cannot be directed at all and is passive, receptive attention. But any movement, even purely mechanical, is vectorial and in that sense directed at something. Here I ignore the fact that intuition is directed in the volitional sense and instinct is not.

13   In *An Introduction to Metaphysics*, Bergson refers to analysis and intuition as 'two ... different ways of knowing a thing' (*Introduction*, 21), which, I assume, means any object including one's own self. Also, one's own 'self which endures' is presented as reality which we undoubtedly grasp in intuition (*Introduction*, 24).

4 Intuition: Knowledge by Acquaintance

From the Bergsonian position, the epistemological faculty that is able to provide us with genuine knowledge must fulfil the following criteria:

1 Like intelligence, it must be able to direct its attention at will.
2 Like instinct, it must embed itself into the object of enquiry and follow it in its entirety – not selecting fragments of it for analysis and arbitrary synthesis.
3 Unlike intelligence, it must not conceptualize, for conceptualization commits one to regard the object in a certain fixed way, whereas real objects are fleeting and changing. Conceptualization also restricts reality to corresponding to concepts whereas real objects are richer in their content than concepts can ever be.
4 Unlike instinct, it must reflect on the object, but not be one with it, otherwise it would disintegrate into a pre-reflective mode of knowledge.

Rational but not conceptual, flexible in the choice of its object and yet rigorous in its grasp of it – this is the type of knowledge that Bergson wants us to see in intuition. Can an epistemological faculty fulfil features so apparently incompatible?

Intuition is aimed at reconciling instinct and intelligence, retaining their strong sides and overcoming the limitations that both instinct and intelligence entail. The strong side of instinct is that it is embedded in being, but it cannot disengage from the immediacy, hence it is restricted to the pre-reflective domain. Intelligence chooses its object at will but it always originates from some practical point of view and has some purpose to its attention, hence the information it accumulates is a selection of raw materials for future action, with many aspects omitted from the picture. Intelligence deals with concepts and, having supplied us with a concept, makes us either seek a reality which would match this concept and leave other features of it out, or rearrange the reality to make it fit into the concept, forcing us to distort reality in either case.

Despite having strong views about the merits of intuition, Bergson is not clear on many issues connected with it. However, our observation that intuition is a fully rational act which is performed at will is confirmed by Lacey and Mullarkey: 'The ways in which intuition goes beyond instinct are that it is conscious, confined to humans, and reflective'.[14] Mullarkey observes that Bergson 'encourages us to "plunge" and "insert our will" into perception, "deepening", "widening" and "expanding" it as we do'.[15] Whenever Bergson talks of intuition he always associates it with an effort to reverse our habitual way of thinking and this in itself involves a rational decision, an act of will. Also, we can rehearse Bergson's own statement made in *Creative Mind*: 'My intuition is reflection' (*CM*, p. 88).

Lacey and Moore describe an act of intuition as holistic: duration cannot be grasped by separate considerations of its sections but 'must be apprehended as a whole, like a melody'.[16] Moore adds: 'Bergson's holism is ... psychological or experiential, rather than logical'.[17] Russell suggests qualifying this type of epistemological approach as synthesis: 'The essential characteristic of intuition is that it does not divide the world into separate things, as the intellect does; although Bergson does not use these words, we might describe it as synthetic rather than analytic'.[18]

One could raise concerns in regard to the claim that intuition grows out of instinct. Bergson's belief in the infallibility of intuition should be founded on a belief in the infallibility of instinct: embedded in the onto-logical structure of being, how can it but *know* the very essence of it? However, observations of instinctive behaviour discredit its alleged infal-libility. Russell observes: 'Instinct, as a rule, is very rough and ready, able to achieve its results under ordinary circumstances, but easily misled by anything unusual'.[19] He draws on examples taken from various sources which illustrate errors in instinctive behaviour: chicks follow any moving

---

14   Lacey, *Bergson*, 150–1.
15   Mullarkey, *Bergson and Philosophy*, 159.
16   Lacey, *Bergson*, 154.
17   Moore, *Bergson: Thinking Backwards*, 42.
18   Russell, *The Philosophy of Bergson*, 10.
19   Bertrand Russell, *The Analysis of Mind* (London: George Allen and Unwin, 1949), 55.

object resembling their mother (James); ants tend the larvae of a beetle, which eats the young of the ants, with the same care that they tend their own (Dr Drever). Also, Bergson himself admits in passing that 'instinct is, like intelligence, fallible' (*CE*, p. 182).

But even if instinct could not be misled by unusual circumstances, it needs to be emphasized that Bergson's epistemological views do not accommodate a theory of knowledge in a strict sense because neither intuition nor instinct involve the capacity of falsehood. In those examples where instinctive behaviour goes wrong, it is not because of a false judgement but rather because of the lack of knowledge, ignorance, which both the chicks and the ants demonstrate. For Bergson who emphasizes the functional, creative power of intuition, true or false statements about an object do not matter. What matters to him is whether the organism or the mind knows what to do with the object, and the genuine insight into the nature of the object consists, for Bergson, in becoming infected with the specific dynamic properties of the object, with the action that is unrolling within it.

As we strive to uphold the idea of subjectivity of the self, we cannot let ourselves be led by Bergson into the direction of accepting cognitive ability merely as a faculty to continue some work of creation, whether as instinctive behaviour or artistic work. We need to look for knowledge as a faculty to appreciate and evaluate data whilst remaining on one's own premises, not contaminated by these data. From this perspective, Bergson's intuition as it is portrayed explicitly in his texts and as it is understood by his critics will not suffice.

5   Three Stages of Cognition

Looking beyond the explicit in Bergson, we find that intuition appears in two modes: as pre-conceptual and as post-conceptual. Pre-conceptual intuition of an object occurs prior to our analysis of it, and post-conceptual intuition is the result of our attention superseding the conceptual framework and reverting back to the original mode of acquaintance enriched, however, by the intermediary conceptual reflection. We claim that the cognitive process in Bergson really involves three stages, which we shall term pre-conceptual (primary) intuition, conceptual intellection and post-conceptual (secondary) intuition.

This is where some secondary literature errs: critics pick up either one or the other stages of intuition, overlooking the fact that there are, in fact, two stages. Kolakowski criticizes intuition as 'incommunicable',[20] reminding us that Bergson describes it as symbol free. He stresses that intuitive insight can only be a private achievement, which seems like a dead end for an epistemological faculty. What Kolakowski misunderstands here is that such intuition is but a beginning of rationality, pre-conceptual intuition.

Herman, on the other hand, thinks of secondary intuition when saying that 'far from rejecting the intellect and its concepts, intuition has need of them to communicate itself'.[21] George Rostrevor asserts that intuition must include reflection,[22] but it is Bachelard who, we think, gets closer to the truth:

> We see the relations between intuition and intellect as more complex than a simple opposition. We see them as constantly co-operating when they come into play. There are intuitions at the root of our concepts: these intuitions are unclear: they are wrongly thought to be natural and rich. There are intuitions too in the way we put concepts together; these essentially secondary intuitions are clearer: they are wrongly thought to be artificial and poor.[23]

Our understanding of the three stages of the epistemic process in Bergsonism is as follows. First, data of consciousness are received in pre-conceptual intuition which resembles Husserl's pre-predicative evidence, included in

---

20   Kolakowski, *Bergson*, 29.
21   Herman, *The Philosophy of Henri Bergson*, 46.
22   Rostrevor, *Bergson and the Future of Philosophy: An Essay on the Scope of Intelligence*, 60–1.
23   Bachelard, *The Dialectic of Duration*, 30–1.

predicative evidence when later conceptualized.[24] Our primary intuition of the worldly phenomena is richer and broader than anything we ever express verbally, and the conceptualized versions of our perceptions are selective and hence always incomplete. We are consciously aware of more than we are able to communicate.

For example, if we wanted to give an exhaustive description of an object, saying 'I see a black ink pen fifteen cm long with an arrow shaped silver logo printed on its side' would be only a start, because we would have to find ways to mention each crevice, each deviation of colour, the size and the degree of fading of the logo, any damage caused by usage, minute scratches on the surface, visible dust particles, dents and marks – in other words, everything that we see and are aware of but what normally does not receive attention and is not conceptualized when we talk about pens. What receives attention when we talk about pens is their colour, whether they work or not, brand and price, but not minute scratches on the surface and the fading of the logo, although these are facts that we consciously observe about pens.

Or we could be looking at an unusual phenomenon and struggle to describe it within the available conceptual framework, but this does not mean that we are not accessing it in a rational and reflective way. This experience could be memorized and conceptualized later using, say, a specialist language.

That which is retained and remembered in later recollections is, of course, an image, and we suggest that it is images that we access intuitively in our perceptions. Bergson himself uses the word intuition to refer to the accessing of images in pure perception: '... instantaneous intuitions, on which our perception of the external world is developed' (*MM*, p. 66). However, he admits that the role of these intuitions is limited as the datum of immediate intuitions is 'a small matter compared with all that memory adds to it' (*MM*, p. 66). I would say, however, that there is no reason why the pre-conceptual intuition we are talking about could not include data

---

24   Edmund Husserl, *Cartesian Meditations: An Introduction to Phenomenology* (London: Kluwer Academic Publishers, 1977), 11.

delivered both by perception and memory jointly. For when we perceive images and our immediate perceptions are accompanied by memory-images, this joint awareness of the immediate data presents itself to our mind as rational and yet pre-conceptual, or that which exceeds conceptuality. We may even suggest that memory-images are necessary for pre-conceptual intuition because they facilitate understanding and account for the element of rationality there.

Kolakowski stresses that 'intuition is supposed to give us direct, yet not sensual, contact with reality, "direct" meaning that it dispenses with abstract concepts'.[25] His remark is useful as it prevents us from identifying intuitive data with pure perceptions. Pre-conceptual intuition takes place after the initial sensual pre-reflective contact but before any conceptualization of the perceived datum, and temporally coincides with the perceived process. As the event unrolls, our mind, in the mode of pre-conceptual intuition, follows it and coincides with each new phase of the process.

It is understandable why Bergson, when trying to illustrate intuition, talks of listening to music. When we listen to music, we often remain at the stage of pre-conceptual rational intuition without moving on to the stage of conceptualization. Music is hard to conceptualize, and for listeners, it remains an undivided flow of sound. A lover of music can, of course, associate particular musical episodes with images like cascades and waves, or concepts like love and joy, and a trained musician is able to identify and name musical notes. This conceptualizing of music, however, which does not happen in all cases, is obviously secondary and obviously superimposed on the initially perceived and consciously acknowledged flow. This makes listening to music a good illustration of intuition which is more than perception, since it is acknowledgement, albeit not yet conceptualization.

The fact that primary intuition is in itself incommunicable is not a problem because it is replaced by concepts very quickly. Carr's observation helps accommodate intuition and intellect in an epistemological act without one excluding another:

25   Kolakowski, *Bergson*, 28.

> So far as activity is an intuition it shares its character of reality with the whole of experience, but the moment it is regarded as a representation of reality which interprets to us the universe, it has ceased to be an intuition, and it cannot be said that the intellect fails to comprehend it. It may fail to harmonize it, but that is another matter.[26]

As a process, pre-conceptual intuition presents an intelligent grasp of the object which precedes rationalization. Pre-conceptual intuition is essentially contemporaneous with the event intuited, and as soon as our mind wishes to stop following the event and ponder on previous stages of the event which may have only just passed, this primary intuition subsides and gives way to intellectual rationalization. However, intuition is not lost forever – Bergson refers to the remainder of it that accompanies an intellectual act as 'a fringe of intuition', as observed by Carr.[27]

According to Bergson, that phase of intuition which we termed preconceptual cannot help but provide a genuine insight into the nature of things, because it consists in attention that follows the perceived process and coincides with it temporally without relapsing into memory or leaping forward into anticipation. We have identified this epistemological accomplishment as knowledge by acquaintance, opposed to the absence of knowledge, as in Russell. Pre-conceptual intuition runs parallel to the perceived process and doubles it, reflects it, in consciousness, so we get reflection without turning to the past. This attention must keep following the perceived process, and thus it has no time to form concepts about it. As soon as concepts emerge, this is a sign that attention has retracted itself from the object immediately given and turned to past memories which aid it to elaborate on the perceived and retained data.

Pre-conceptual intuition does not constitute statements or judgements about the perceived processes. It is a prolongation, an extension of this process. The intuition that I have of the event is part of that event, with the event emanating, giving away an image that evokes perception in me. In intuition I am not an outside observer; I am involved in the all-embracing ontological process of being and acknowledgement of being.

26 Carr, 'Bergson's Theory of Knowledge', 57.
27 Carr, 'Bergson's Theory of Knowledge', 60.

In pre-conceptual intuition one is a passive recipient of excitations which are received via the senses and registered in the mind. Bergson insists that this knowledge provides an insight into what is essential in the object, and it is in this sense that he claims this knowledge to be genuine. Inevitably, intuition crystallizes itself in concepts, which are needed for communication, it is true, but not just for conveying messages to others. We need concepts to help us understand and come to terms with what we have intuited. The being, of which our intuitions are part, needs to be interpreted. We need to communicate with ourselves as well as with others, and data received in intuition needs processing and evaluating.

Why do we need this? Should intuitive insight not be enough, at least for our own satisfaction? The answer is no. We remember from our study of *Matter and Memory* that the self needs the materiality of the immediate situation to anchor all its phenomena onto one central point. The self needs definition; without it, we become disorientated as in daydreaming or, even more so, in our sleep. To define ourselves we constantly verify where we are, when and in what circumstances, by referring to immediate data and coordinating them with our memories and plans. To communicate thus with our own selves we need to be able to have a mental grip on the objects of our experience, memories and plans. If all our psychic events, including past memories, contribute to the nature of our self, we need to access them if we want to understand what we are. Concepts serve as cues that evoke relevant images, and also substantialize our relations with the world, giving names to times, dates, locations and our roles – the process which is evaluated by Bergson as secondary to immediate knowledge, but necessary for our sanity.

Is the conceptualized picture of the world a false picture, according to Bergson? Not quite. When he criticizes rationalized knowledge, he does not claim it to be false in the sense that it affirms something which is the reverse of the truth about the real. This is not as erroneous as mistaking white for black, but it is a translation of genuine data into practically useful symbols which are meant to represent reality but which misrepresent it instead. The symbol reflects that side of the object which comes into contact with us in our interactions with the world, ignoring its other sides. The role of objects as our tools and commodities (or obstacles for our actions) is emphasized to the point that it overshadows anything else in the objects and is presented as their essence. We, however, mistakenly believe that what symbols represent is the essence of things. For Bergson, the temporality and the being of reality are irreducible, hence everything in the real is essential. Pre-conceptual intuition reflects a far wider scope of facts than a conceptual account, and, whereas it does not capture the event in its absolute entirety, it is the best source of initial data merely due to the amount of information grasped at this stage.

After conceptualization, we end up in a virtual world of things and events supposedly set out to revolve around men and serve them, the things' essence derived from their referential relation to human action. Intelligence modifies the data we receive about the real in order to fit it into the anthropocentric world, ascribing to beings and things the view that they are *in themselves* and *for themselves* essentially what is encoded within the role that they play in interactions with us.

Intelligent approach means a shift of emphasis from decentred to anthropocentric, to incomplete and one-sided but not essentially false data. Most of the time, and in line with our human nature, we remain at this level of intelligent interpretation of the world, wrapping it around our needs. Only sometimes, according to Bergson, do we demonstrate ability for what we term post-conceptual intuition. Whereas pre-conceptual intuition is absorbed in the fleeting moment and ignores other reality, post-conceptual intuition is far from playing the part of a passive receptacle for everything that takes place.

Čapek is another Bergsonian commentator who offers a one-sided view of intuition. He ignores primary intuition, but his account provides a useful insight into secondary intuition. Čapek's explanations concern the phase in which intuition supersedes intelligence. Intuition, he says,

> begins with the attitude of distrust for the accepted modes of thought; ... This attitude of distrust originates in a vague awareness of certain experiences incompatible with the accepted modes of thought; the essence of intuition is precisely to bring these vague and rather implicitly felt data into a clear focus and to show that the new forms of understanding thus created are superior to the old ones.[28]

---

28   Čapek, *Bergson and Modern Physics*, 97.

Their superiority, Čapek explains, consists in a greater flexibility, an ability to explain more aspects of experience than the previous forms of knowledge, as well as accounting for data already acknowledged and interpreted.

In Čapek's interpretation, intuition (post-conceptual intuition, let us remember)

1. Is a complex process;
2. Has nothing in common with emotion and instinct;
3. Does not go on effortlessly and passively;
4. Beginning as a hazy anticipatory feeling, it is transformed into the clarity of understanding, with its content being purged of all parasitic elements. What was previously clear, appears as an oversimplification of the data of experience.

It is the post-conceptual intuition that lies, we think, behind intuition as a philosophical method and intuition as artistic creation, for neither a method nor artistic activity can be imagined at the level of the pre-conceptual approach to reality. Pre-conceptual, primary intuition is too primitive to be held responsible for an intellectual insight. On the other hand, secondary intuition, which supersedes concepts, can accommodate them and intuit their multiplicity as something more than just the sum of units. It sees beyond the immediacy of each individual concept and observes another kind of immediacy – the immediacy of the concealed connections amongst concepts and amongst things.

To illustrate that post-conceptual intuition builds on concepts, we can recall that in *An Introduction to Metaphysics* Bergson talks about an intuitive insight of a writer who researches his subject and makes numerous notes first, and then, by a 'painful effort', places himself or herself 'at the heart of the subject'. The writer then achieves an impulse which,

> once received, starts the mind on a path where it rediscovers all the information it had collected, and a thousand other details besides; it develops and analyses itself into terms which could be enumerated indefinitely. The farther we go, the more terms we discover; we shall never say all that could be said, and yet, if we turn back suddenly upon the impulse that we feel behind us, and try to seize it, it is gone; for it was not a thing, but the direction of a movement, and though indefinitely extensible, it is infinitely simple. (*An Introduction*, p. 61)

Whereas pre-conceptual intuition seems to involve a passive dissolving of one's mind in the world, post-conceptual intuition is active and assertive as it forms part of a creative process, whether it be writing a novel or painting. Just like pre-conceptual intuition, secondary intuition involves knowledge by acquaintance and cannot be either true or false. It is engaged in creative work, and is bound with the new reality that it itself initiates. Just as in pre-conceptual intuition, there is no judgement or statement which could be true or false: knowledge here helps to create being, it does not represent being and does not conform to the contemporary idea of what is conventionally understood by cognition.

Normally, pre-conceptual intuition gives way to conceptual acknowledgement of the object which, as Bergson observes, facilitates our interactions with the external world. The aim of conceptual knowledge is to detach itself from things and create symbols, which point at things and statements, which are about things, and which in themselves do not contribute to the structure of things. We could say, though, that conceptual knowledge, just like instinct and pre-conceptual intuition, is part of being in general because it forms part of our interactions with the external world. Thus conceptual knowledge gives us partial, verifiable by experience, propositional and communicable truths about reality, understood in terms of our interactions with it: these truths are necessarily anthropocentric.

Most of the time our epistemological activity remains within the framework of conceptual knowledge. It can, however, be superseded by post-conceptual intuition. This happens when a large collection of concepts gives rise to a new insight which in turn dismisses the concepts as it has created a new vision of things, which is manifested in a creative act. Post-conceptual intuition is part of a creative process, just as instinct and pre-conceptual intuition are part of being. The difference is that in post-conceptual intuition, the creative subject stops being a passive recipient of data but instead he himself or she herself directs the creative process. Just like primary intuition, secondary intuition is incommunicable, and must be conceptualized (for example in a critical review of the work of art) if the content of this knowledge needs to be communicated.

Looking at epistemic processes that penetrate the hierarchy of being, we find the foundation of any knowledge on the level of basic physicality, in physical proximity, which can develop as physical contact. On the biological level, there is instinct which is ontologically bound with being and cannot be either true or false. The psychological level of cognition includes three stages: pre-conceptual intuition, which is, like instinct, ontologically bound with being, cannot be either true or false and is incommunicable; conceptual knowledge, which has referential relation to being and comprises true and false statements about it, although these truths are only partial and anthropocentric; post-conceptual intuition, another form of cognition that cannot deliver propositional truths because it is ontologically bound with being in a creative process.

As well as questioning the validity of knowledge from the position of its referential certainty, we can look at how it functions as part of being. From this perspective, instinct is a directing force in biological life as it directs growth and biological behaviour. Pre-conceptual intuition is an engaging force which effects the simultaneous following of things by the mind. Conceptual knowledge is a mediating force between various types of being where intelligence binds people and things. Post-conceptual intuition is a directing force in the creation of radically new man-made reality.

Bergson's epistemology opens up an attractive possibility of an all-inclusive exploration of the self: if any knowledge has, one way or another, an ontological value, then so must the knowledge about one's own self have it. Thus we do not have to establish truths which hide behind the beliefs we have about our self and which are separate from these beliefs, and failing to find those truths, dismiss the beliefs as false and artificial constructions. One way or another, all data that we accumulate through introspection in themselves constitute our self so we do not have to attempt to treat these data as mere symbols of what the self is. The way forward, deduced from Bergsonism, is to accept that introspective data at least represent themselves if nothing else, and not to search beyond the obvious, but to accept the obvious for what it is. Our guiding principle in this will be taking for granted the fact that no thoughts or feelings we have of our 'I' should be dismissed as illusions, because even illusions must be taken for what they are, as part of our psychological reality.

6   The Self in Introspection

In everyday language, 'I' refers to the entire eventual diversity of the human being, material, biological, and mental, infinitely compressing all concurrent processes as well as the entire remembered and unremembered personal history into a singularity with the highest possible density of content but the smallest in terms of its peceivability. The search for the 'I' in introspection makes us recede mentally into the depth of our own consciousness, abstracting ourselves from extensity in favour of higher and higher intensity, assuming that what corresponds to the concept 'I' is the ultimate author of our words and actions, the most alert and most clear part of our consciousness.

However, 'I' may mean different things. When we say, 'I am tall', 'I am warm' and 'I am sad', we are not referring to this ultimate focal point, but to ourselves as either a material object, a biological object, or an historical being. 'I am tall', for instance, describes my height – my physical size, a property of every extended body, not at all something that can be a characteristic of a compressed singularity; here we refer to ourselves, to our 'I', as a material extended object.

Of course, 'I' in 'I am tall' does not signify an inert body. Rather it demonstrates that 'I' involves extensity and size, for 'tall' here is a characteristic of one's self and not merely of some physical body to which the self is somehow attached. Equally, 'I am warm' discloses the biological component of the 'I', and 'I am sad' indicates that 'I' is a rational historical process, with memories of the past which, combined with current eventuality, result in a concrete sentiment.

In our everyday understanding of the 'I', the conviction of it being the intense core of consciousness co-exists with our using 'I' to refer to our basic physicality and our biology, inasmuch as our 'I' cannot escape or disengage in any real way from our size and our sensations. In our attempt to find, to pinpoint, the 'I' inside us, we condense, we concentrate all of our being into a nucleus which must equal the 'I am' in its most compact form but which must also include our body and our history. We make everything in our being gravitate inwards, not divorcing our material, biological and historic properties but compressing them further and further, despatializing them in the hope of grasping the irreducible phenomenon of our conscious 'I'.

How does the idea of 'I' enter our mind? The rational self is an historical self, an historical process, acknowledged and evaluated. Its identity as one is effected by memory holding together temporally diverse phases of the self, so self-consciousness necessarily involves looking back and comparing the self now and the self at other times. But does the idea of 'I' emerge from within one's own being, or does it require social interactions with other individuals?

Suppose the inner movement of biologically derived feelings could supersede itself by becoming a generalized acknowledgement of all pre-reflective processes of selfhood. This process of generalization would also be a process of infinite compression of the diverse psychological content in the past, present and possible future into one idea – that of the 'I'. The 'I' would be the product of consciousness – a kind of a super dense nucleus self-forming out of the fuzzy nebula. Could this process, and the idea of 'I' as a result of this process, occur by itself? The Bergsonian answer, we believe, would be negative.

According to Bergson, the brain and, as we understand it, consciousness cannot create anything new: whatever the sentient being produces is a modification of the movement existing previously and coming from elsewhere (*MM*, p. 73).[29] So, consciousness could originate within one's psyche only if it were an ontological prolongation of the pre-reflective processes, if rationality could originate from the sensuality alone, and if consciousness could grow out of biology.

As we understand Bergson's theory, we can assign the pre-reflective sphere and all processes that precede it to the domain of the instinctive,

---

29   An alternative reading of Bergson, founded on other texts, is possible here: hetero-
     geneity continuously creates something new and this novelty which comes from
     within duration supersedes causality as genuine creation (see *CE*, 172–3). Also when
     a free act is compared to artistic creation one can understand this as an indication
     that Bergson talks of genuine novelty that 'spring[s] from our whole personality'
     (*TFW*, 172) rather than being incited from outside.

and rationality and the reflective self to the domain of intelligence. We remember Bergson stressing that in the evolutionary movement, intelligence is not a prolongation of instinct but is an independent faculty (*CE*, p. 142). Therefore Bergson would not accept the derivation of the rational self from the pre-reflective self. For him, we guess, pre-reflective self would not be prolonged, by some autonomous self-development, into the rational self; the latter is a qualitatively different psychological event, a different process.

There is a lot of passivity in the pre-reflective self: processes such as sensations and feelings happen because they cannot help but happen. Self-consciousness, on the other hand, is an effort in which the unity of the self turns its attention to itself; instead of looking outwards, it looks inwards. Thus self-consciousness is an effort which the pre-reflective self would not be capable of by itself and this seems to be almost a 'stop and rewind' of the process that we observe in the pre-reflective self, where sensations and moods pass through each other. In self-consciousness the processes do not just pass – they are consciously acknowledged, and the experiences of them are deepened and intensified: if the pre-reflective self allows feelings to expire, the reflective self keeps hold of them. The pre-reflective self involves memory that memorizes and recognizes, but the reflective self uses memory that remembers, recalls and refuses to forget; the memory of the pre-reflective self goes with the flow, but the memory of self-consciousness swims against the tide. It would go against Bergsonism to accept a process with a tendency to reverse upon itself; this tendency, we assume, should come from an outside stimulus.

Emmanuel Levinas may be of help here because he spells out clearly what Bergson merely implies. Levinas recognizes the initial passivity of the pre-reflective self as well as the need for the outside interference in order for the self to become self-conscious. He demonstrates that the pre-reflective processes of selfhood become reversed in the presence of another person. Another person's presence in itself has the effect of a demand or, as Levinas puts it, of an accusation.[30] In response to this accusation one is

---

30   Emmanuel Levinas, *Otherwise than Being or Beyond Essence*, transl. Alphonso Lingis (Dordrecht, Boston, London: Kluwer Academic Publishers, 1997), 113.

forced to turn one's attention in search of the accused, and identifies the accused as 'self'. Levinas stresses that our first encounter with our self occurs through the eyes of another person, and therefore is primarily identified in its accusative state as 'me' rather than as 'I'. 'Everything is from the start in the accusative. Such is the exceptional condition or unconditionality of the self, the signification of the pronoun self for which our Latin grammars themselves know no nominative form'.[31] And: 'The word *I* means *here I am*, answering for everything and for everyone'.[32]

Levinas constructs self-consciousness out of an ethical impulse, ascribing to the latter the divine status of that which exists 'beyond essence'. Without sharing Levinas' excessive preoccupation with ethics, Bergson offers a view on social relations which, not unlike that of Levinas, draws on the primary importance of one's responsibility towards others. Our first memory, Bergson observes, is that of obeying a prohibition that constitutes our first social encounter.

> The remembrance of forbidden fruit is the earliest thing in the memory of each of us, as it is in that of mankind. … What a childhood we should have had if only we had been left to do as we pleased! … But all of a sudden an obstacle arose, neither visible nor tangible: a prohibition. Why did we obey? The question hardly occurred to us. (*TSMR*, p. 9)

It may be correct to assume that self-consciousness in Bergson has a social origin, and that the concept 'I' is formed by the response to the outside influence when the child is addressed by other people. Bergson links intelligence and sociability in humans, and this also indicates the link between self-consciousness and sociality. Effort and concepts indicate intelligence, unlike an effortless and indefinable instinctive act. In self-consciousness there is conscious effort to retain and acknowledge the processes of selfhood and the concentration of them in the concept 'I'. In *Time and Free Will*, we remember, Bergson talks of the role that acts of intellect, conceptualizing acts, play in our social life: 'An inner life with well distinguished moments

---

31   Levinas, *Otherwise than Being or Beyond Essence*, 112.
32   Levinas, *Otherwise than Being or Beyond Essence*, 114.

and with clearly characterized states will answer better the requirements of social life' (*TFW*, p. 139). He explains that social interactions require this rationalization and conceptualization of reality, and so self-consciousness, the belief in one's definite 'I' must also go hand in hand with social interactions, inasmuch as 'I' is a concept.

Let us say that our pre-reflective self becomes faced with the presence of another conscious human being, and we are forced to consider ourselves from the position of another person, who addresses our person as unity. One then looks for that unity, for what corresponds to the 'you' in the demand. Of course, the social stimulus would not be the only factor that turns the pre-reflective self into the self-conscious self. But if we ask whether the pre-reflective self alone could develop into the rational self in Bergsonism, which would account for the continuity of these two stages of selfhood, we can now say that continuity cannot be established due to the necessary involvement of the social demand, and this reveals yet another discontinuity in the Bergsonian duration.

'Me' is the view we take of ourselves which is imposed on us by others and integrated into the stream of our consciousness as 'I', the concept which presents to us the ultimate limit of introspection. Looking inside ourselves, we fail to find what exactly matches the concept 'I', because the concept 'I' has not merely emerged as a signifier that named some piece of reality waiting to be named. The idea of one's selfhood emerged in social interactions, and in itself became a new reality which did not exist on the biological level. The conceptual view of oneself, the idea of our 'I' itself, is new reality existing in its own right.

The conceptualization of one's self is a process that continues the development of psychological duration and is necessary for our nature as gregarious beings (we can only communicate with each other as definite entities, as centres of selfhood), even though Bergson sees it as a mistakenly distorted view: 'As the self thus refracted, and thereby broken to pieces, is much better adapted to the requirements of social life in general and language in particular, consciousness prefers it, and gradually loses sight of the fundamental self' (*TFW*, p. 128).

Bergson says that the true self consisting in 'a confused multiplicity' of psychic states is distorted by language which represents it as 'discrete multiplicity' of psychic states named, separated and juxtaposed (*TFW*, p. 129). But as we accept every manifestation of the self to be an addition to its existence as duration, then even those elements that, as Bergson states, misrepresent the self in self-consciousness, are nevertheless part of the self as its content. We could say that what becomes misrepresented by the concept 'I' is the fuzzy pre-reflective self, but this conceptualized misrepresentation is in fact a new addition to our being made by self-consciousness, which largely (albeit not exclusively, of course) consists in ascribing concepts to one's own being.

It could be argued that Bergson does not need a central agency – an 'ego' or 'I'– to give consciousness an identity. Such identity is already ensured by the fact that moving, embodied, durational consciousness is particular, and its new content is always specific and concrete.[33] Reflective 'I' would be for Bergson a secondary form of individuation. However, we still need to include it in the duration of one's consciousness – as a component of our duration that cannot be eliminated from it.

The main question in this chapter was whether rational introspection provides reliable data about self-consciousness. Since Bergson favours intuition and opposes it to rational conceptual intellection, the revealing of the self in self-consciousness appears to be a distortion of the real self. But if we rely on intuition for introspective data, we face a dead end, because intuition is not communicable even to oneself. On the other hand, conceptual intellection is part of psychological duration because psychological duration comprises all inner psychological events and rational thinking is one of these events. Consequently, we decide to accept data of introspection, but with the following reservations. There is a conflict between two of Bergson's implicit claims: the claim that all psychological events are included in the irreducible psychological duration, and exclusion of conceptual thought and rationality in general from psychological duration.

33 Bergson's theory of duration does not deny individuality and personality. In 'The Problem of Personality' he talks of 'a single and identical person' (Mélanges, 1055), asserting that '[p]ersonality ... is ... a continuity of change' (Mélanges, 1063). Also, it is '[t]his continuity of change causes what we call the permanence, the unity, the substantiality of the person' (Mélanges, 1064).

Further development of the Bergsonian thought requires reconciliation between the above statements. We find that the first one is stronger than the second, and that the products of rational thought should be included in psychological duration as a variety of psychological events. Also we take into account another Bergsonian claim that data, information, is in itself a type of being, and not merely a reference to some other, proper being. From this we conclude that the introspective rational datum should be accepted as phenomenon per se, without necessarily questioning its relation to other parts of the self which it may or may not reflect. In other words, concepts that we form in relation to ourselves exist as legitimate components of the self.[34]

An analysis of Bergson's treatment of concepts can be found in *Mullarkey* (1999), 152–5 and in *Mullarkey* (1997), 44–58, where it is emphasized that although Bergson is opposed to using concepts representing a fixed referent, he introduces a novel use of metaphors which are part of the same process-reality as the process that they refer to, which indicates that the treatment of language and communication in Bergson does not amount to a straightforward rejection of conceptuality. Bergson's including creative emotion into his theory of language sophisticates the issue even further: when a piece of creative writing is inspired by an emotion, and contaminates the reader with that emotion, it guides intellect and conceptuality in an act of intuition (*TSMR*, pp. 45–6).

---

34 Of course, this does not mean that just because we can form certain concepts, they have any legitimacy as concepts. They are genuine as phenomena, but not necessarily in any other sense.

CHAPTER 9

Evolutionary, Historical and Biographical Continuity

1 The New Within Continuity: Staggered Temporal Reality

In this chapter we shall show that the theory of constructive retrospectivity, incorporated in the concept of heterogeneous duration in Chapter 5, is necessary for the idea of continuity in evolution, history, and biography. Despite our most sympathetic attitude to Bergson, we found that discontinuity and discord disrupt the continuity of selfhood at different levels. In its pre-history we observe leaps from matter to biological organisms and then to consciousness; the existence of a concrete self is marked by various inner splits in self-perception, memory and volitions; social existence involves confrontation of selves and by no means harmony and continuity. These discords can be qualified in more general terms as the failure to account for new elements within continuity. We face a reality in which each new temporal phase contains a new additional step added on to the whole. Bergson sees no problem with this, as he is ready to equate all reality with time, explaining new reality by the fact that it corresponds to the new stretches of time. However we, albeit accepting that time is an essential component of reality, refuse to treat it as totally equal to reality. Thus we cannot get away from the fact that in each new temporal phase we witness elements which do not seem to be founded on the previous phase.

Bergson is satisfied with the view of temporal reality as portrayed in Diagram 7, where the future is non-existent and uncertain, and the past is the source of the present: immense and inexhaustible, it increases all the time and out of its own depths produces the present, which is only a thin layer of reality.

Future

Present

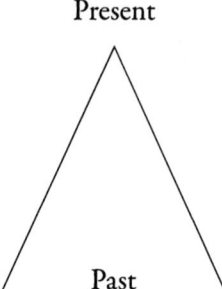

Past

Diagram 7   Bergson's view of temporal reality

We disagree with this view. The picture of the temporal reality that we accept is presented in Diagram 8.

Future

Present

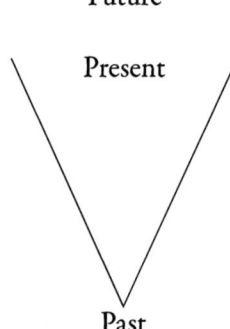

Past

Diagram 8   Our view of temporal reality

Diagram 8 shows that, since the reality enlarges, its previous content is poorer than its following content, so its past has poorer content than its present, and its more remote past has poorer content than its more recent past. The reason why Bergson accepts an image of the infinitely vast past is that he ignores the issue of temporal relations. He does not differentiate between earlier and later and the more or less remote past.[1]

---

1   See Chapter 6 for a discussion on temporal relations.

Treating temporal relations as ontologically non-essential would break up all temporal processes, because each phase of a process can be divided indefinitely into shorter and shorter phases, and if their phases are no longer in a strict sequence coordinated by the relations of before and after, then they can be in any sequence. We would argue that the past has inherent temporal structure. If *x* happened before *y*, then in whatever way the past is preserved, this relation is preserved with it: 'earlier' and 'later' are irreducible ontological characteristics of temporal phases. Replacing the Bergsonian picture with a structured sequence, we can see that the following temporal stretches are richer in content than the preceding ones, because they inherit a legacy from the preceding ones and also contain new elements. The present is richer, 'larger' than the past, and this larger, new content includes something that the previous phase could not fit in. This expanding reality cannot be sufficiently explained by its past.

It could be said that the picture I have sketched of Bergson's relatively undifferentiated past ignores his discussion of planes of the past in *Matter and Memory*, and illustrated by Bergson's own diagram:

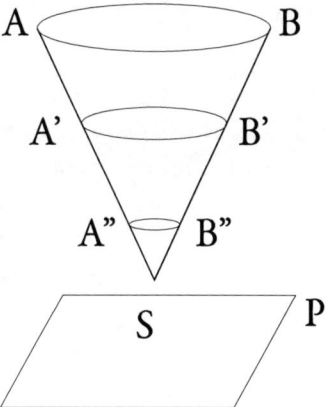

Diagram 9 Planes of the past (See *MM*, pp. 152 and 162)

Cone SAB The totality of the recollections accumulated in one's memory.
Base AB Motionless; situated in the past.
Summit S Indicates one's present, moves forward and touches the plane P.
Plane P One's actual representation of the universe, changing all time.
Sections A'B'; A"B" Repetitions of one's psychical life.

[B]etween the sensori-motor mechanisms figured by the point S and the totality of the memories disposed in AB there is room ... for a thousand repetitions of our psychical life, figured by as many sections A'B', A"B", etc., of the same cone. We tend to scatter ourselves over AB in the measure that we detach ourselves from our sensory and motor state to live the life of dreams; we tend to concentrate ourselves in S in the measure that we attach ourselves more firmly to the present reality, responding by motor reactions to sensory stimulation. In point of fact, the normal self never stays in either of these extreme positions; it moves between them, adopts in turn the positions corresponding to the intermediate sections, or, in other words, gives to its representations just enough image and just enough idea for them to be able to lend useful aid to the present action. (*MM*, pp. 162–3)

However, as can be seen from the above, Bergon's planes of the past do not seem to correspond to a temporal ordering within past reality and do not refer to more or less remote slices. What they refer to is a degree in which our psychic life concerns itself with either pure remembering or embodied practical existence. As Worms indicates, this schema refers to psychological life (but not to the temporal ordering of reality) and illustrates the double movement of the body at S (situated in the plane P of images) and memories descending from AB towards S. The planes, which are closer to the base of the cone, correspond to deeper planes of memory (but not necessarily, as we understand it, to a more remote past) and the planes closer to the top of the cone correspond to memories which are loaded with perception.[2] Those memories that are mixed in with perceptions are not necessarily most recent ones but have been drawn from the depth of memory by the needs of the current situation. As for the pure, unremembered, past it apparently remains undivided and un-sliced.

If we consider reality at different times during its unrolling – as we approach time on a daily basis, when referring to a stretch of time already gone – we will see it crudely staggered as in Diagram 10. If we compare temporal slices t, t' and t" we can see the lack of continuity between the past and the present, or the previous and the following periods of time, as this shows the abruptness with which new content is attached to the already existing duration.

2   Worms, *Introduction à Matière et Mémoire de Bergson*, 320–1.

Future

Present

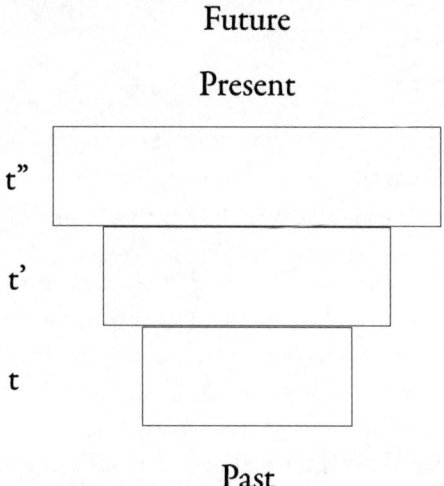

t"

t'

t

Past

Diagram 10   Temporal slicing of reality reveals leaps between earlier and later

The above analysis confirms that supposedly continuous duration contains leaps and abrupt changes within itself. But if we appeal to the intuitive perception of duration from within – say, of our own life – we regain the feeling of continuity whereby new events flow out of the previous ones. Why does this feeling of continuity contradict the abrupt picture of reality that our metaphysical examination reveals? In the next section we will claim that the problem that we discovered lies in fact in the observer's position (bringing about spatialization of reality, which even the metaphysical approach cannot avoid) and will confirm that the perception of temporal continuity as continuity is possible when one is literally immersed in its temporality and shares its history. That confirms Bergson's claim that the true picture of duration is only achievable from within.

2   The Observer's Position

When discussing temporal continuity and finding leaps within it, we overlooked the fact that such a discussion involves an observer – the one who
imagines and discusses it. In this we followed an implicit assumption that
we could in principle adopt a view of reality according to which it unrolls
from the past to the present and beyond. But the observer's position, itself
part of temporal reality, gives a certain perspective of what we observe, relative to our position, and affects the way we perceive things.[3] This, in turn,
indicates that the problems we encountered may be due to this position
and not necessarily inherent in the observed reality.

In our analysis of continuity, together with Bergson, we adopted an
imaginary position where we observed temporal processes from the most
remote past, as in Diagram 11.

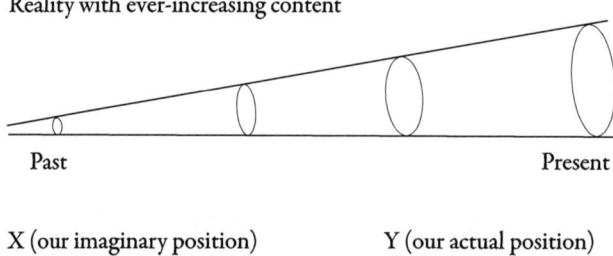

Reality with ever-increasing content

Past   Present

X (our imaginary position)   Y (our actual position)

Diagram 11   Temporal reality augmenting as time passes, and our observation positions

By positioning ourselves in this way we gain a projection of reality from
the past to the present, with its later phases being greater in content than
its earlier phases. What follows cannot equal that which precedes it as it
does not contain only that which precedes, and so we find ourselves unable
to explain the present by the past. Following the order in which events

---

3   A discussion of the relativity of the observer's position in space-time can be found
     in Alexander, *Space, Time and Deity: The Gifford Lectures at Glasgow 1916–1918*, 77.

happened seems to be the most logical and natural way to view reality, but in actual fact, we do not follow this order. In our imagination, we abandon our real position Y and place ourselves in the remote past X, treating that past as the present observation point, but we actually still belong to our present, which is the ultimate point of the unrolling reality that we observe. Thus we end up with an overlap of two observation points: the point of our embodied existence in the actual present (Y) and the point in the remote past (X) which we treat as the present.

When we do not try to be positioned in two different presents simultaneously, and remain in the actual present, we observe the unrolling reality of which we are part and find each new phase consistent with the old one, immersed in the continuity of temporal existence around us. When we observe events whilst living them, there is nothing in them that seems to appear out of nowhere, as if by magic, but when we throw ourselves back into the remote past and attempt to follow the track of events from the remote past to the actual present, we stumble against abrupt changes. Our point of view is split between the actual present and the remote present back in the past, and both positions interfere with each other. When we remain in our actual present, we do not know what the next temporal phase will be, so we do not focus on searching for the foundations of that next phase. In the situation of the split position, the imaginary observer, thrown back in time, is informed by the actual observer in the present of the sequence of events which followed that remote position in time. Thus, if we agree to treat that imaginary position as present, the future of that present is fixed for the observer, something that Bergson strongly disagrees with. Indeed, when that past was present, its future was uncertain, therefore there could not have been a solid foundation for that future because it finalized itself only when it became present.

As discussed in Chapter 5, consequent events affect the nature of the preceding events, transforming them into the foundation for the following events. So when, as the present observer, in our imagination we position ourselves 'at the back of time', we deal with the past which has been transformed by the following events, and not with the unscathed temporal reality which was unscathed only when it was present. As the observer of past events, we insist on treating the remote past as if it were both the unscathed present, unaffected by the future, and the past foundation of the following events, affected by those events which were future but have finalized themselves as present at some stage in the past. As follows from Chapter 5, when the present reality becomes past, it is not the same reality any more: it is altered by its relations with the subsequent events.

It would have been possible to have had a view of the remote past as the intact present if we could have disengaged from our own present and transferred to that time whilst it was present. Then our current present would not interfere with the past which we are trying to comprehend. But then we would not have been able to set ourselves the task of looking for the past foundation of present things, and our aim of trying to establish continuity of the old and the new still could not have been achieved. In each case we would get a view relative to the chosen position, and not gain access to the way things really happened. We assert that temporal reality centred on the earliest imaginable present is not the ultimate ontological truth but merely a possible view of temporal reality, and that we can have different views depending on the actual or imaginary position of the observer.

Now if, instead of placing ourselves behind reality and looking forward, towards our actual position, we remained where we are and looked backwards, we would not see leaps from old states to new states with richer content: instead of the ever-increasing complexity of reality we would see the ever-decreasing complexity. This view will more readily give us a sense of continuity because instead of the addition of new elements we will observe their elimination, and descent from the richer to the poorer content, which is far easier to comprehend than the ascent from the poorer to the richer.

A critic will say that the suggested alternative to the view from the past is absurd because it goes against the natural course of events: things do not move backwards in time. But we demonstrated in Chapter 5 a way in which the present changes the past, whereby following events throw back threads of relations which affect preceding events, so that changes in the present cause certain changes in the past. These latter changes happen backwards in time, so when we follow the reality from the remote past to the actual present, we miss these changes. By missing them, we miss continuity because the temporal reality then presents itself to us as a ladder with each new step containing elements unfounded on previous steps. If we let ourselves look backwards in time, from the actual present to the past, we see that each new step had thrown its light on the preceding step and retrospectively made it its foundation. Before the following step emerged, the preceding step could not have contained a foundation for the following step, because the following step, being future, was not definite at that stage until it became present. It could have had a different content, and then the preceding step would have become a foundation for a different following step.

As temporal reality changes, some or all new elements of it are not continuous with the previous state if looked at from the point of view of the previous state which by itself is unable to take into account the consequent state. Whilst present, consequent events are future in relation to it, and we accepted that the present cannot have a definite relation with the future because the future itself is indefinite whilst it is future. In order to form a relation with the real consequent event, the consequent event must define itself by becoming present, but this means that the formerly present state will have to become past. From the position of the new present state, the past state is its immediate foundation and the source of its being but the relation that maintains the continuity does not originate from the previous state but from the following state which emanates a causal link with the immediate past and makes it its own predecessor.

3   The Continuity of Evolution

Although we distinguish physical, biological and psychological processes of selfhood, they are elements of the heterogeneous duration of selfhood and do not exist separately from one another. In Chapter 7 I emphasized that the highest, conscious processes of selfhood pierce through all three levels,[4] and so we widened the scope of the Bergsonian heterogeneity: deriving heterogeneity from the psychological self, Bergson discussed only

4   See Chapter 7, Sections 4 and 5.

heterogeneous psychic states, whereas we, applying the notion of hetero-geneity back to the self, included in it everything that constitutes the real self, from the mineral base of human bones to ideas of the human mind.

Treating mind and body as heterogeneous qualities of duration, we still need to account for their unity. If our observation that temporal reality develops backwards as well as forwards is correct, then we may be able to comprehend the continuity of the worldly development from basic matter to the human soul, but in the reverse order. We assume now that the rela-tion that ensures the continuity of cause and effect does not radiate from the cause but from the effect, because by the time the effect took place, the event which has become the cause is no longer present.

In the attempt to comprehend continuity, we can adopt a view allow-ing us to see the causal links between temporal stages which are made ret-rospectively, remaining where we are, in the actual present, and following temporal reality back from the present towards the past. The embodied present is not purely present reality which belongs exclusively to the present, did not exist in the past and will not exist in the future. The appearance of the present reality alone is an indication that there was the past which must have been an immediate foundation for this present, and that there will be the future because what we witness now is unlikely to disappear when this present is over. The immediate past, in its turn, requires its own immediate past and so on indefinitely.

In fact, we ordinarily move backwards in time, because an analysis of something present entails a survey of its components, and the pre-existence of the components is ontologically prior to the existence of the whole. When we observe an object, we see its present state and its history encom-passed in that present state. When we want to unroll that history, we natu-rally work backwards from the present state to the past and not vice versa, because our starting point must be something obvious, and that can only be the present state available for perception. The remote past is something we arrive at after constructing a continuity that takes us backwards.

The present state, when deciphered, gradually reveals the past to us, and in a present perception of a person we acquire his or her latest perfor-mance, combined with the legacy of his or her personal history and the preceding history of evolution.[5] The dynamic view of the person that we gradually get is not that of a progression from minerals to the soul, i.e. from the embodied past to the embodied present, but a regression from the soul to the minerals, i.e. from the present to the past. Looking backwards, we see that the soul depends on the biological and the physical base even though, attempting to work from the past to the present, we fail to see how the soul grows out of the physical base. If we start from the beginning of time, we can never derive the soul from the minerals, but by looking at the entire person, we can see that his or her conscious existence comprises biological and physical existence, and whereas it is not possible to observe in reality the ascent from minerals to the soul, we can observe the opposite movement in the disintegration of a living person into a biological structure maintained by machines and then to mineralized remains.

If we accept that our usual way of tracking reality from the remote past to the present is merely a possible, not ultimate view, we may as well choose a different position and follow the history of its being not from the past to the present, but in the opposite direction, the direction of descent, starting from the soul, not finishing with it. Then we get a sense of unbroken continuity, where thoughts and desires are dependent on one's bodily states and structures, and which in their turn are founded on basic inert physicality, all this being a condensation of the previous evolutionary movement and of the total development of being, from its first appearance to forms of the highest complexity.

Recalling the Bergsonian notion of a qualitative multiplicity with diverse qualities brought together in a dynamic unity, we will emphasize the qualifier 'diverse'. For the qualities that constitute the human being are diverse to the extreme, combining inanimate matter, living matter, thoughts and feelings. Although all these qualities exist simultaneously and concurrently in a concrete human being, in terms of the history of the worldly development, some appeared and established themselves earlier

---

5   Contemporary biology does not equate evolution with progress and complication of structures and, contrary to Bergson, does not regard humans as an apex of being in evolutionary terms.

and some later on the scene of being. The human being thus represents a materialized, compressed record of the preceding worldly development, and a diachrony of inanimate matter, living matter, and mind, where each component can account for itself and for the previous, simpler stages, but not for the following, more complex stages.[6]

Thus the bond between soul and body can originate only from the soul because the soul is in principle temporally later than a biological body – not in any concrete sentient being but in general. We remember that although in *Matter and Memory* Bergson equates spirit and matter, insisting that neither is derivative of the other (*MM*, p. 9), he does not dispute the dependence that spirit may have on matter, and shows in *Creative Evolution* that sophisticated mental and emotional activity depends on the sophistication of the biological base, which demonstrates that ultimately, life comes before reason, and that mind, or soul, depends on the body.

Since souls cannot exist without bodies but bodies can live without souls, the soul can be considered the effect that turns the body into its cause, and we need to look into the soul for the explanation of the mind/body problem. Actually, this is what has been happening all along: attempts to understand the mind/body issue have always been initiated by a conscious, not bodily, effort, and the search has always been defined by conscious means of collecting and processing data, not by some bodily effort. Possibly, the difficulty here (as it has been noted by Bergson in his criticism of intellect) lies in the fact that consciousness, wrapped in its own interests and ways of being, such as logic and analytical rationality, remains in this elevated realm, failing to descend to the body, reluctant or unable to consider the body in bodily terms.

The solution may lie in Bergsonian intuition, i.e. a cognitive faculty that supposedly combines the rationality of consciousness and the straightforwardness of a bodily function, regarded as the descent from the heights of the soul into the lower levels of biology and physics. Interestingly, Mark Johnson offers a theory of the cognitive faculty which underlies all our

6   For a demonstration of the involvement of preceding states in higher stages, where mind comprises surrogates of inanimate and living matter, see Chapters 7 and 8.

rational negotiations with the world and which is based on our being a physical object and a living body. He talks of 'nonpropositional aware- ness' which we have, say, of our 'being balanced upright in space', 'even though all my efforts to communicate its reality ... will involve proposi- tional structures'.[7] Johnson claims that rationality in general is founded on such nonpropositional knowledge and maintains that 'the conceptual/ propositional content of an utterance ... is possible only by virtue of a complex web of nonpropositional schematic structures that emerge from our bodily experience'.[8]

According to Johnson, our pre-linguistic bodily involvement in the world creates models which we use when constructing our representation of the outer world. The modes of bodily existence, it being a container, a moving object, an agent of force, produce schematic structures which we apply to worldly phenomena and to abstract ideas. As an example, we will describe Johnson's schema of balance and show its evolution from direct bodily experience to an abstract notion.

Firstly, we learn balancing with our body as an activity and not by adopting rules and concepts. We know the meaning of balance through experiencing body equilibrium, or lack of it. Apart from standing upright, balance is conceived (still pre-conceptually) as a harmonious functioning of one's organism, with just enough food in the stomach, the right body temperature etc. We learn the lack of balance if our hands are too cold, or we have eaten too much. The sense of bodily balance, having produced a sche- matic structure of experience, is then applied to visual perception. Works of art are perceived as balanced if their elements produce a visual equilibrium of space occupancy, imaginary weight and mass. When we see an object against some background, we sense immediately whether it is harmoniously positioned, or whether it is off centre. The same schema applies when we talk of psychological states. Psychological balance, Johnson observes, is understood as equilibrium between bodily and mental states. When this correlation is out of balance, one can be overwhelmed by emotions or be

7   Johnson, 4.
8   Johnson, 5.

emotionally drained, and the balance is restorable either by releasing the built up emotions or generating emotional energy. In a similar way the schema of balance is applicable to rational arguments, justice and mathematical equality, where the concept of equally distributed weight is used in a metaphorical sense.

Nonpropositional awareness does not answer the question: how did body produce consciousness during the course of evolution and how does body maintain consciousness in a concrete person? Johnson's theory does not explain in full the ontological dependence of consciousness on the body but offers a convincing picture of the way in which knowledge both as a process and as content is guided by bodily performance. If nothing else, Johnson's theory illustrates how physics and biology guide rationality, and this provides evidence of at least partial continuity of body and mind, where the body determines not the fact of existence of the mind but the mode of its existence, and where this dependence is not manifested by the body but exclusively by the mind.

4   Historical and Biographical Continuity

As well as participating in life and being in general, a concrete self comprises the embodied time of its historical existence as a social being. The social history of the self, we claim, can be compared with the historical continuity of a state rather than an organism: the unity of the self harbours splits and confrontation between its different parts, resembling subjects of a historical process.

Historical events gain their identity when they are complete and past. Going back to the First World War example,[9] the assassination of the Archduke in itself does not entail the necessity of the war, and cannot sufficiently explicate it as its consequence. When it took place, it was uncer-

9   See Chapter 5, Section 3.

tain whether Serbia as a state was to be made accountable for the crime. The Austrian side could have decided to treat the murder as an act of lunacy and one could imagine possible reasons for doing so. When the assassination took place and the ultimatum to Serbia was issued, Russia may not have announced its passing to the 'Period Preparatory to War' so promptly, or at least the Russians could have chosen a different wording which would have had less aggressive connotation and reflected better the position of the Russian government whereby this move was meant to be a 'purely precautionary measure'.[10] If Austria had understood the Russian position as purely precautionary, there may not have been German and Austrian mobilization in response, etc, etc. At any time the powers involved could have made a different move and, if the war was still unavoidable, it would have had a different identity. It is only after the chain of events leading to the war took place, that we identify them as the prelude of *that* First World War.

We can define the meaning of an event only retrospectively, when all stages of this event have finalized themselves by becoming present and the entire event is therefore in the past. For whilst the event is still unrolling, at each stage it may change its direction, and eventually have different content and different consequences for the events which will succeed it in the future. This is especially evident when we are involved in critical events which at the outset indicate significant changes for our lives, whose outcome we cannot predict for certain: will we pass that examination, marry that person, get that job? There is even less certainty when we live through events that affect an entire society.

Someone living through a revolution, for example, would be overwhelmed by the sense of instability, rapid political changes, rising prices and, disorientated, would find it hard to project his or her own future. When the crisis is over, all its phases and episodes become part of the revolution, but they were not part of the revolution before because, while they were happening, it was uncertain what the entire event would become. It may not have become a revolution but could have prolonged itself as a civil war

---

10   Everett, *World War I*, 17.

or it could have subsided into an economical recession with limited, not radical political changes. Then the episodes that happened, say, at the outset, would have become members of a different heterogeneity and themselves would be assigned a different content.

The same can be said about our personal history, whereby we live our latest present with the sense that all previous events have continuously led us to this moment. But from the point of view of ourselves, say, ten years younger, if we had known then what we would be doing now, we would have been astonished, for the change might have seemed very abrupt and our past life might not have seemed to show any predispositions for this later development.

Suppose, at time t one is introduced to a group of colleagues, and three years later, at time t' he marries one of them. Now, if at time t he were told about what was to happen in three years time, he would not find it easy to accept this piece of information. To him it may have almost seemed like an act of violence, an invasion of a different reality into his actual life, having to accept a complete stranger as his closest relation. On the other hand, three years later, such a thought does not seem odd at all, and he refers to the first meeting with the new colleague as the first meeting with his future wife. At time t, however, he would be absolutely right to reject a hypothetical prophesy of their marriage. Such prophecy would not be legitimate because at time t his new acquaintance was not his future wife since, for a multitude of reasons, their relationship may have not developed in the future. It is only in times consecutive to time t that their liaison arose and strengthened, culminating in marriage at time t'. Now, whereas from the present position at time t, this marriage would seem an unlikely accident, from the present time t', when the marriage is a finalized event, the chain of causal links radiating from this event towards the past time t and beyond, obviously form a continuity leading to the marital union.

5 The Hypothetical Temporality of the Self

The future of a historical process appears to be greatly uncertain. Rather than being guided by remote future goals, political figures are pressurized by the demands of immediate problems, and have little, if any, room for manoeuvre whilst making historical decisions. Also, history involves wills of millions combined with unexpected objective circumstances such as a natural disaster or a discovery of a new source of oil, so it is hard to plan and even harder to predict. Hence history is synonymous with the past of social reality, and historians dare not make authoritative predictions of its future.

A historical process of the self is more straightforward because, given all its inner splits, it ultimately involves only one willing agent, so an individual has a greater control over his or her personal future than subjects of world history have over the future of history. Our actions are not a mere response to circumstances – we make plans and work on their realization.

This complicates the temporality of the self in the following way. We concluded in Chapter 5 that with each new present, the entire world or, in the case of a concrete self, the self's life changes, and these changes include alterations of the past and of the future. Now we must say that as well as being influenced by the past, the self is also influenced by the future – not by the future of real events because they are not certain yet, but by the future which we ourselves invent; not by the future which is projected by the past and the present and which constantly changes, but by the hypothetical, imagined future which is a state that we strive to achieve, which will not happen by itself as a natural continuation of our previous life but requires a special, purposeful effort.

Thus in our present we negotiate two futures, the real projection of our present and the hypothetical future of our goals. Whereas the real future emanates from the present and is viewed from the present point, the relation of the hypothetical future to the present is the reverse. When we act in our present with the view of the hypothetical future, as an acting agent we remain in the present, but as an observing and monitoring agent we place ourselves at the point of the planned achievement in the hypothetical future and evaluate our actual present from that position. Our imaginary position in the future is treated as the imaginary present, and from the point of view of this present, our real present becomes the past of our imaginary present.

Viewing our actual present from the position of the distant goal, we gain a retrospective view of it and modify it from the foundation of the real future into the foundation of the hypothetical future, with the aim of changing the hypothetical future into the real future. Then the present also appears in two ways. Firstly, it remains that actual present which is a spindle that turns temporal reality and changes the past and the future, but secondly it, whilst being looked at retrospectively, gains for us the properties of a reality that is affected by subsequent times and subsequent events. The current day and the current hour project themselves naturally into tomorrow and into the next hour, but they are also seen as potentially the foundation for the goal in the more remote future.

The importance of the hypothetical future is especially evident in our relation with death. Typically, the knowledge of its inevitability does not cause anxiety, unless we believe that we know the time of our final hour. The reason why knowing makes such a dramatic difference to our state of mind is that the not knowing means uncertainty not only epistemological but also ontological, given the uncertainty of the future in general, and presents us with a potentially indefinite life.

The knowledge of the time of one's death is more destructive than the event of death itself, because when expected, death affects the hypothetical temporality in which we create our plans which can stretch beyond our real temporal limits. We start to suffer from death mentally and emotionally before we are dead biologically, because our temporal plane of action is crudely cut ahead of us, rendering us unable to live in the hypothetical future. The trauma of such awareness is equally great whether we are promised ten years of life or ten days. Someone who will accidentally die the next day is luckier than someone who is sure of dying in ten years time.

However, human beings are sometimes capable of overcoming even the knowledge of death. Wills and testimonies, legacies and living gifts allow the self to emanate its will beyond the cut-off date. An interesting case of self-therapy when facing death can be found in Boethius' *The Consolation of Philosophy*, written whilst the author was awaiting execution.[11] Book 5, Prose 6 of *Consolation* contains the view of eternity, which opens up a possibility for the temporal reality to supersede its temporal immediacy in an unlimited fashion.[12] The core argument there is that God in one moment grasps all events in the past, present and future as non-ceasing. The act of embracing God in faith allows the believer to partake to some extent of this vision of time and eternity and thus, ignoring the knowledge of the coming end, indulge in the temporal fullness of being. Whilst alive, we are able to think of God, and think of the all-encompassing reality, overcoming in our mind the thought of our end. Boethius' psychological exercise, we think, is an intensification of the normal human tendency to reach out in our thoughts to all possible times outside our own.

As for our belonging to the actual present time, it is effortless in physical terms, but requires an effort when we want to belong to the social present time. The social time is marked by cultural phenomena, including the material culture, and fashion awareness can be a means of belonging to the social present, which has a different duration for different people. Some may be happy to belong to a more extended present and use elements of the material culture which are ten or fifteen years old, whereas others regard the present to be a shorter, three to five year long period. But there are those who are satisfied only if they belong to a present as narrow as possible, lasting only for a few months. In truth, fashion, car and machinery industries often do not offer anything functionally different in the latest versions of their goods and what they are really selling is the pass to the ultimate, cutting-edge social present, as they understand it. On the other hand, there are people who disapprove of the newfangled ways and things, and escape into the past, but both groups consider concrete temporality in axiological terms, with the embodied 'earlier' and 'later' entailing characteristics of better or worse.

11   For a further analysis of Boethius' understanding of eternity see Eleonore Stump and Norman Kretzmann, 'Eternity', *The Journal of Philosophy*, Vol. 78, No. 8 (August 1981), 429–58.

12   'Eternity ... is the complete, simultaneous and perfect possession of everlasting life' (A. M. S. Boethius, *The Consolation of Philosophy*, transl. V. E. Watts (London: Penguin, 1969), 163).

To complicate our temporality even further, we have to account for the time before our birth and for the time after our death, i.e. the time, both past and future, that we cannot possibly grasp experientially for ourselves. Also, we often follow a fictional or historical narrative, when reading a book or watching a film, which portrays events as unrolling now but which are not happening now in reality. Interestingly, such layering of an imaginary present on top of one's own present is not problematic and does not entail confusion of one's identity. The real person, however engrossed he or she may be in a story, remains in his or her real present, and the superficial identification with an imagined character can play a therapeutic role giving, so to speak, a vacation from the pressures of the real time.

Apart from various forms of linear temporal belonging, there is a cyclical, or spiral mode of temporal reality, where the content of some events is essentially the same as that of previous events. Routine, daily and yearly, also defines our temporality, regulating our sense of stability. In this context, the repetitiveness and the originality in the routine are also regarded in axiological terms: routine can be judged positively as tried and tested stability, or negatively, associated with boredom and monotony, and originality positively as an improvement, or negatively as unproven and unreliable innovation.

Conclusion

This enquiry, focused on Bergson's philosophy of time, passed through three stages. Firstly, I followed the conceptual development of duration – the key concept that applies to all types of temporal reality. The goal of this stage was to clarify and discipline Bergson's discourse of time so as firstly, to make Bergson's theory of time more accessible for a non-specialist reader and secondly, to demonstrate that, despite his worthy contribution to ontology and epistemology, Bergson's theory is incomplete and naturally requires further development.

The second stage consisted of a thorough analysis of the metaphysical terms which emerged from the examination of Bergson's texts – duration, heterogeneity, time, past, present, future. The theory of heterogeneity which Bergson merely outlined has received further development. I analysed elements of this heterogeneity and concluded that the identity and individuality of its elements, albeit not detectable ostensibly, can be identified via *Heterogenels* their effects on the world. The nought has been reinstated in ontology: Bergson denies nothingness any ontological value, treating it as a quality, but I suggested that it participates in heterogeneity as a relation of otherness. It has also been argued that as everything affects everything else in heterogeneity, each new addition in the present will rearrange the entire whole, including the projected future and the past, and this changes the future prospects and gives new meanings to historical episodes. Finally, I explained the unavoidable failure to capture the very present of events by a temporal shift between the body which goes through the present phase of a physical action, and the conscious acknowledgement which, in its own present, can either anticipate or remember the present of the body but not coincide with it temporally. *Coincide with the temporality of the body*

As for time and temporal parts, I argued that Bergson was wrong in reducing time and everything temporal to qualities, because time involves relations as well: temporal processes objectively happen earlier and later than other processes. I also demonstrated that the past, present and the future can be understood in either qualitative or relational terms: in qualitative terms, if taken as temporal reality in general, and in relational, if treated as past, present and future.

The third stage involved concretizing the idea of duration. It was decided that an all-embracing duration is best illustrated by the self because the self could either entail or account for the full diversity of being. Analysing the structure of rational selfhood, we modified Bergson's theory of cognition, removing the opposition of intuition and analysis and showing that the cognitive process involves both intuition and analysis as stages of cognition. I also found that on different occasions Bergson refers to two different types of intuition, and suggested a three-staged epistemology consisting in primary (pre-conceptual) intuition, intellectual analysis, and secondary (post-conceptual) intuition.

I found that the duration of selfhood is peppered with various types of discontinuity, which seems to contradict the idea of duration. An attempt to restore continuity in duration was made in the final chapter, where I suggested an alternative view on temporal reality – not trying to follow it from the past to the present, but from the present to the past, whereby we avoid the overlap of the actual point of view in the actual present and the imaginary point of view in the past which we treat as if it were present.

As was pointed out at the outset, I do not insist that Bergsonism should develop necessarily in the direction indicated in this study. But I insist that if we want to allow it to exist, it *should* develop, because Bergson's own theory is clearly unfinished. What makes it worthy of the philosophical attention is its potential as a system of concepts and insights that expose those aspects of reality, which may remain overlooked otherwise. In particular, as well as bringing in the dynamic dimension into being, Bergson's theory of duration affirms the heterogeneous nature of reality, which, I believe, has its merits, despite also being problematic. Whilst offering an all-inclusive theory of temporal being which includes knowledge, Bergson unites ontology and epistemology. Also, his idea of heterogeneity, where the identity of a fragment of temporal process is not exhaustible by its ostensive characteristics but involves its relations with other fragments and with the whole, prompts a new theory of identity. I hope that this project has succeeded in bringing Bergson's originality to light and showing at least one way in which the theory of duration can be developed further.

T. V. Orlovskaya and I. V. Mirenkova, *Duration* (2012)

Bibliography

Bibliographical Note

Time and Free Will (1889), which introduces the idea of duration in connection with psychology and motion, was Begson's doctoral thesis. It was followed by *Matter and Memory* (1896), a complex analysis of matter and consciousness. *Creative Evolution* (1907), a study of duration manifested as life drive, as creation of life, became immensely popular soon after its publication. *Mind-Energy* (1919) is a collection of essays on the mind-body problem. *Duration and Simultaneity* (1922) contains a discussion with Einstein on the theory of relativity. *The Two Sources of Morality and Religion* (1932) applies the theory of duration and intuition to social issues. Bergson's last book *The Creative Mind* (1934) is a collection of various essays.

For an extensive bibliography of books and articles which deal with Bergson's thought, see J. Alexander Gunn, *Bergson and His Philosophy* (Whitefish, Montana: Kessinger Publishing, 2004) 104–59.

Bergson's texts

Bergson, Henri, *Creative Evolution*, transl. Arthur Mitchell (London: Macmillan, 1964) (*CE*).
——, *The Creative Mind*, transl. Mabelle L. Andison (New York: Citadel Press, 1974) (*CM*).
——, *The Creative Mind*, transl. Mabelle L. Andison (New York: Philosophical Library, 1946) (*CM* [hardback]).
——, *Duration and Simultaneity*, transl. Leon Jacobson (Manchester: Clinamen Press, 1999) (*DS*).
——, *An Introduction to Metaphysics*, transl. T. E. Hulme (Indianapolis, Cambridge: Hackett, 1999) (*Introduction*).

——, *Matter and Memory*, transl. Nancy Margaret Paul and W. Scott Palmer (New York: Zone Books, 1991) (*MM*).

——, *Mind-Energy: Lectures and Essays*, transl. H. W. Carr (Westport: Greenwood Press, 1975) (*ME*).

——, *Time and Free Will: An Essay on the Immediate Data of Consciousness*, transl. F. L. Pogson (London: George Allen and Unwin, 1910) (*TFW*).

——, *The Two Sources of Morality and Religion*, transl. R. Ashley Audra and Cloudesley Brereton (Notre Dame, Indiana: University of Notre Dame Press, 1986) (*TSMR*).

The above texts can also be found in Bergson, Henri, *Key Writings* (New York, London: Continuum, 2002), which also contains extracts from Mélanges, transl. Melissa McMahon.

Original French Publications

Bergson, Henri, *L'évolution créatrice* (Paris: Presses Universitaires de France, 1907).

——, *La pensée et le mouvant* (Paris: Presses Universitaires de France, 1934).

——, *Essai sur les données immédiates de la conscience* (Paris: Presses Universitaires de France, 1889).

——, *Durée et simultanéité.* (Paris: Presses Universitaires de France, 1922).

——, 'Introduction à la métaphysique', *Revue de Métaphysique et de Morale* (January 1903).

——, *Mélanges* (Paris: Presses Universitaires de France 1972).

——, *L'énergie spirituelle* (Paris: Presses Universitaires de France, 1919).

——, *Matière et mémoire* (Paris: Presses Universitaires de France, 1896).

——, *Les deux sources de la morale et de la religion* (Paris: Presses Universitaires de France, 1932).

***

Alexander, Ian W., *Bergson: Philosopher of Reflection* (London: Bowes and Bowes, 1957).

Alexander, S., *Space, Time and Deity: The Gifford Lectures at Glasgow 1916–1918*, Vol. 1 (London: Macmillan, 1966).

Andrade, Jackie, ed., *Working Memory in Perspective* (Hove, UK: Psychology Press UK, 2003).

Antliff, Mark, *Inventing Bergson: Cultural Politics and the Parisian Avant-Garde* (Princeton, New Jersey: Princeton University Press, 1957).

Aquinas, Thomas, 'Treatise on the Angels', Thomas Aquinas, *The 'Summa Theologica' of St Thomas Aquinas* (New York: T. Baker, 1911).

Augustine, St, *Confessions*, transl. R. S. Pine-Coffin (Harmondsworth: Penguin, 1961), Book XI.

Ayer, A. J. *The Central Questions of Philosophy* (London: Penguin, 1990).

Bachelard, Gaston, *The Dialectic of Duration*, transl. Mary McAllester Jones (Manchester: Clinamen Press, 2000).

Baddeley, Alan, *The Psychology of Memory* (New York, Basic Books, 1976).

Baddeley, Alan and Wilson, Barbara, 'Amnesia, Autobiographical Memory, and Confabulation' in David C. Rubin, ed., *Autobiographical Memory* (Cambridge: Cambridge University Press, 1988), 225–52.

Bakhurst, David, 'Lessons from Ilyenkov', *The Communication Review*, Vol. 1, No. 2 (1995), 155–78.

Barnouw, Erik, *Documentary: A History of the Non-fiction Film* (Oxford: Oxford University Press, 1993).

Barsalou, Lawrence W., 'Flexibility, Structure, and Linguistic Vagary in Concepts: Manifestations of a Compositional System of Perceptual Symbols', in Alan F. Collins et al., eds, *Theories of Memory* (Hove, Hillsdale: Lawrence Erlbaum Associates, Publishers, 1993), pp. 29–102.

Bennington, Geoffrey, *Lyotard: Writing the Event* (Manchester: Manchester University Press, 1988).

——, 'Time after Time', *Journal of the British Society for Phenomenology*, Vol. 32, No. 3 (October 2001), 300–11.

Boethius, A. M. S., *The Consolation of Philosophy*, transl. V. E. Watts (London: Penguin, 1969).

Bradley, F. H., *Appearance and Reality: A Metaphysical Essay* (Oxford: Clarendon Press, 1978).

Bréhier, Émile, 'Images Plotiniennes, Images Bergsoniennes', *Les Études Bergsoniennes*, Vol. 2 (1949), 105–28.

Brentano, Franz, *Philosophical Investigations on Space, Time and the Continuum*, transl. Barry Smith (London, New York: Croom Helm, 1988).

Brewer, W. F. and Pani, J. R., 'The structure of Human Memory', in G. H. Bower, ed., *The Psychology of Learning and Motivation: Advances in Research and Theory* (New York: Academic Press, 1983), Vol. 17, 1–38.

Butters, Nelson and Laird S. Cermak, 'A Case of Study of the Forgetting of Autobiographical Knowledge: Implications for the Study of Retrograde Amnesia', in David C. Rubin, ed., *Autobiographical Memory* (Cambridge: Cambridge University Press), 253–72.

Calhoun, C. and Solomon, Robert, eds, *What is an Emotion?* (New York: Oxford University Press, 1984).

Čapek, Milič, *Bergson and Modern Physics* (Dordrecht: D. Reidel, 1971).

Cariou, Marie, *Bergson et Bachelard* (Paris: Presses Universitaires de France, 1995).

——, 'Bergson: the Keyboard of Forgetting', in John Mullarkey, ed., *The New Bergson* (Manchester and New York: Manchester University Press, 2000), 99–117.

Carr, H. Wildon, *Henri Bergson: The Philosophy of Change* (London: T. C. & E. C. Jack, reprinted by Kessinger Publishing, 2005).

——, 'Bergson's Theory of Knowledge', *Proceedings of the Aristotelian Society*, Vol. 9 (1908–1909), 41–60.

——, 'Bergson's Theory of Instinct', *Proceedings of the Aristotelian Society*, Vol. 10 (1910), 93–114.

Casey, Edward S., 'Habitual Body and Memory in Merleau-Ponty', *Man and World*, Vol. 17 (1984), 278–97.

——, 'Remembering Resumed: Pursuing Buddhism and Phenomenology in Practice', in Janet Gyatso, ed., *In the Mirror of Memory: Reflections on Mindfulness and Remembrance in Indian and Tibetan Buddhism* (New York: SUNY Press, 1992).

Chabot, Pascal, *La Philosophy de Simondon* (Paris: Vrin, 2003).

Collingwood, R. G., *An Essay on Metaphysics* (Oxford: Clarendon Press, 1969).

Coxon, A. H., *The Fragments of Parmenides* (Netherlands: Assen, 1986).

Crocker, Stephen, 'The Past is to Time What the Idea is to Thought or, What is General in the Past in General?', *Journal of the British Society for Phenomenology*, Vol. 35, No. 1 (January 2004), pp. 42–53.

Crovitz, Herbert F., 'Loss and Recovery of Autobiographical Memory after Head Injury', in David C. Rubin, ed., *Autobiographical Memory* (Cambridge: Cambridge University Press, 1988), 273–89.

Crowther, Paul, *Philosophy after Postmodernism: Civilized Values and the Scope of Knowledge* (London and New York: Routledge, 2003).

Crystal, David, *The Cambridge Encyclopaedia of Language*, Second Edition (Cambridge: Cambridge University Press, 1997).

Cumming, Robert, *Annotated Art* (London, New York, Stuttgart: Dorling Kindersley, 1995)

Cunningham, Gustavus Watts, *A Study in the Philosophy of Bergson* (New York: Longmans, Green and Co., 1916).

Deleuze, Gilles, *Bergsonism* (New York: Zone Books, 1991).

——, 'Bergson's Conception of Difference', in John Mullarkey, ed., *The New Bergson* (Manchester: Manchester University Press, 2000), 42–65.

Descartes, R., *Discourse on Method and the Meditations* (London: Penguin, 1971).

Dummett, Michael, 'Bringing about the Past', in Robin Le Poidevin and Murray MacBeath, ed., *The Philosophy of Time* (Oxford: Oxford University Press, 1993), 117–33.

Dywan, J. and Bowers, K., 'The Use of Hypnosis to Enhance Recall', *Science*, 222, 184–5.

Everett, Susanne, *World War I* (London: Hamlyn, 1980).

Forbes, Graeme, 'Time, Event, and Modality', in Robin Le Poidevin and Murray MacBeath, ed., *The Philosophy of Time* (Oxford: Oxford University Press, 1993), 84–95.

Frankfurt, Harry, 'Freedom of the Will and the Concept of a Person', in Robert Kane, ed., *Free Will* (Oxford: Blackwell Publishing, 2001), 127–44.

Gale, R. M., ed., *The Philosophy of Time* (London and Melbourne: McMillan, 1968).

Goethe, Johann Wolfgang Von, *Faust* (Oxford: Oxford University Press, 1998).

Gramer, John G., 'The Plane of the Present and the New Transactional Paradigm of Time', in Robin Durie, ed., *Time and the Instant: Essays on the Physics and Philosophy of Time* (Manchester: Clinamen Press, 2000), 183–6.

Grünbaum, Adolf, 'Time, Irreversible Processes, and the Physical Status of Becoming', in J. J. C. Smart, ed., *Problems of Space and Time* (New York: Macmillan, 1968), 397–416.

Haeckel, Ernst Heinrich Phillip, *The Evolution of Man*, Vols 1 and 2 (Whitefish, Montana: Kessinger Publishing, 2004).

Hegel, G. W. F., *Lectures on the Philosophy of Religion* (Berkeley, Los Angeles, London: University of California Press, 1988).

——, *Hegel's Science of Logic*, transl. W. H. Johnston and L. G. Struthers, Vol. 1 (London: George Allen and Unwin, 1929).

——, *Science of Logic*, transl. A. V. Miller (New York: Humanity Books, 1998).

Heidegger, Martin, *Poetry, Language, Thought*, transl. Albert Hofstadter (New York, Evanston, San Francisco, London: Harper & Row, 1971).

——, *Being and Time*, transl. John Macquarrie and Edward Robinson (Oxford: Basil Blackwell, 1980).

——, *Basic Problems of Phenomenology* (Bloomington, Indiana: Indiana University Press, 1982).

Heisig, James W., *Philosophers of Nothingness* (Honolulu: University of Hawaii Press, 2002).

Henderson, John, *Memory and Forgetting* (London and New York: Routledge, 1999).

Herman, Daniel J., *The Philosophy of Henri Bergson* (Washington: University Press of America, 1980).

Holmes, Richard, *Wellington The Iron Duke* (London: HarperCollins, 2002).

Hornsby, Jennifer, *Actions* (London: Routledge & Kegan Paul, 1980).

Hume, D., *A Treatise of Human Nature*, L. A. Selby-Bigge, ed. (Oxford: Clarendon Press, 1960).

Husserl, Edmund, *The Idea of Phenomenology*, transl. William P. Alston and George Nakhnikian (The Hague: Martinus Nijhof, 1964).

——, *Ideas: General Introduction to Pure Phenomenology* (London: Collier-Macmillan, 1967).

——, *Logical Investigations*, transl. J. N. Findlay, Vol. 2 (London: Routledge & Kegan Paul, 1970).

——, *Cartesian Meditations: An Introduction to Phenomenology* (London: Kluwer Academic Publishers, 1977).

——, 'Pure Phenomenology, Its Method and Its Field of Investigation', Husserl, E.: *Shorter Works*. Peter McCormick and Frederick A. Elliston, eds (Notre Dame, Indiana: University of Notre Dame Press, 1981).

——, *On The Phenomenology of the Consciousness of Internal Time (1893–1917)*, transl. John Barnett Brough (Dordrecht /Boston /London: Kluwer Academic Publishers, 1991).

Huxley, Julian, *Evolution: A Modern Synthesis* (London: George Allen and Unwin, 1942).

Ilyenkov, E. V., 'The Concept of the Ideal', *Philosophy in the USSR: Problems of Dialectical Materialism* (Moscow: Progress, 1977).

James, William, 'Bradley or Bergson?', *Journal of Philosophy*, 7, No. 2 (January 1910), 29–33.

——, chapter 24 'Instinct', *Principles of Psychology*, Vol. 2 (New York: Courier Dover, 1950), 383–441.

Johnson, Mark, *The Body in the Mind: The Bodily Basis of Meaning, Imagination, and Reason* (Chicago: The University of Chicago Press, 1990).

Kant, I., *Critique of Pure Reason*, transl. Norman Kemp Smith (London: Macmillan, 1961).

Kolakowski, Leszek, *Bergson* (Oxford: Oxford University Press, 1985).

Lacey, A. R., *Bergson* (London and New York: Routledge, 1993).

Levinas, Emmanuel, *Otherwise than Being or Beyond Essence*, transl. Alphonso Lingis (Dordrecht, Boston, London: Kluwer Academic Publishers, 1997).

Lewandowsky, Stephan, Dunn, John C., Kirsner, Kim, eds, *Implicit Memory* (Hillsdale, New Jersey: Lawrence Erlbaum Associates, 1989).

Lewis, David, 'The Paradoxes of Time Travel', in Robin Le Poidevin and Murray MacBeath, eds, *The Philosophy of Time* (Oxford: Oxford University Press, 1993), 134–46.

Loizou, Andros, *The Reality of Time* (Aldershot, UK: Gower, 1986).

——, *Time, Embodiment and the Self* (Aldershot, UK: Ashgate, 2000).

Lowe, E. J., 'Tense and Persisitence', in Robin Le Poidevin, ed., *An Offprint of Questions of Time and Tense* (Oxford: Clarendon Press, 1998), pp. 43–59.

Lynn, Steven Jay and Kevin M. McConkey, eds, *Truth in Memory* (New York: Guilford Press, 1998).

Lyons, W., *Emotion* (Cambridge: Cambridge University Press, 1980).

Marietti, Angèle, *Les Formes du Mouvement chez Bergson* (Paris: Les Cahiers du Nouvel Humanisme, 1953).

McTaggart, John McTaggart Ellis, *The Nature of Existence*, Vol. 2 (Cambridge: Cambridge at the University Press, 1968), Chapter 33, 'Time', 9–31.

Mellor, D. H., *Real Time II* (London: Routledge, 1998).

Merleau-Ponty, Maurice, *Phenomenology of Perception*, transl. Colin Smith (London: Methuen, 1965).

Moore, F. C. T., *Bergson: Thinking Backwards* (Cambridge: Cambridge University Press, 1996).

Moore, G. E., *Philosophical Studies* (London: Routledge & Kegan Paul, 1965).

Mulhall, Stephen, *On Being in the World. Wittgenstein and Heidegger on Seeing Aspects* (London and New York: Routledge, 1993).

Mullarkey, John, 'Bergson and the Language of Process', *Process Studies*, Vol. 24 (Spring 1997), 44–58.

——, *Bergson and Philosophy* (Edinburgh: Edinburgh University Press, 1999).

——, 'Introduction: La Philosophie Nouvelle, or Change in Philosophy', John Mullarkey, ed., *The New Bergson* (Manchester: Manchester University Press, 2000).

——, ed., *The New Bergson* (Manchester: Manchester University Press, 2000).

——, 'Forget the Virtual', *Continental Philosophy Review*, Vol. 37 (2005), 469–93.

Murphy, Howard 'Australian Aboriginal Concepts of Time', in Kristen Lippikoff, ed., *The Story of Time* (London: Merrell Holberton, 1999), 264–7.

Parfit, Derek, 'Later Selves and Moral Principles', in A. Montefiore, *Philosophy and Personal Relations*, ed. (London: Routledge & Kegan Paul, 1973).

——, *Reasons and Persons* (Oxford Clarendon Press, 1991).

Pears, D. F., 'Time, Truth and Inference' in Anthony Flew, ed., *Essays in Conceptual Analysis* (London: Macmillan, 1966), Chapter 11, 228–4.

——, ed., *Freedom and the Will* (London: Macmillan, 1965).

Pearson, Keith Ansell, *Philosophy and the Adventure of the Virtual: Bergson and the Time of Life* (London: Routledge, 2002).

Piaget, J., *Play, Dreams and Imitation* (New York: Norton, 1962).

Poidevin, Robin Le and Murray MacBeath, eds, *The Philosophy of Time* (Oxford: Oxford University Press, 1993).

Plato, *Sophist* (Whitefish, Montana: Kessinger Publishing, 2004).

Plotinus, *The Enneads*, transl. Stephen MacKenna (London: Faber and Faber, 1969).

Price, H. H., *Perception* (London: Methuen, 1932).

Rice, T. Talbot, *Icons* (London: Studio Editions, 1990).

Rostrevor, George, *Bergson and the Future of Philosophy: An Essay on the Scope of Intelligence* (London: Macmillan, 1921).

Roy, Edouard Le, *A New Philosophy: Henri Bergson* (Whitefish, Montana: Kessinger, 2004).

Russell, Bertrand, *The Philosophy of Bergson* (Cambridge: Bowes and Bowes, 1914).

——, *The Analysis of Mind* (London: George Allen and Unwin, 1949).

——, *The Problems of Philosophy* (Oxford: Oxford University Press, 1983).

——, *Human Knowledge: Its Scope and Limits* (London: Routledge, 1992).

Ryle, Gilbert, *The Concept of Mind* (London: Hutchinson & Co., 1969).

Sait, Una Bernard, *The Ethical Implications of Bergson's Philosophy* (Whitefish, Montana: Kessinger Publishing, 2005).

Salmon, W. C., ed.: *Zeno's Paradoxes* (Indianapolis, Indiana and New York: Bobbs-Merrill, 1970).

Sartre, Jean-Paul, *Being and Nothingness: An Essay on Phenomenological Ontology*, transl. Hazel E. Barnes (London: Methuen, 1969).

——, 'Imagination and Emotions', in Jean Paul Sartre, *Jean-Paul Sartre: Basic Writings* (London: Routledge, 2001), 89–105.

Schacter, Daniel L., *Searching for Memory – the Brain, the Mind, and the Past* (New York: Basic Books, 1996).

Schank, R. C. and Abelson, R. P., *Scripts, Plans, Goals and Understanding* (Hillsdale, New Jersey: Erlbaum, 1977).

Scharff, Robert C., Dusek, Val, ed., *Philosophy of Technology* (Oxford: Blackwell Publishing, 2002).

Schopenhauer, Arthur, *The World as Will and Representation*, transl. E. F. J. Payne, Vol. 1 (New York: Dover Publications, 1969).

Schumann-Hengsteler, Ruth, Martin Strobl, and Christof Zoelch, 'Temporal Memory for Locations: On the Coding of Spatiotemporal Information in Children and Adults', in Gary L. Allen, ed., *Human Spatial Memory* (Hove, Hillsdale: Lawrence Erlbaum Associates, 2004).

Skinner, B. F., *Science and Human Behaviour* (New York: Macmillan, 1963).

Smart, J. J. C., 'Mind and Brain', in Richard Warner and Tadeusz Szubka, eds, *The Mind-Body Problem: A Guide to the Current Debate* (Oxford: Blackwell, 1997).

Stephen, Karin, *Misuse of Mind: A Study of Bergson's Attack on Intellectualism* (London: Routledge, 2001).

Strawson, Galen, 'The Self', in Shaun Gallagher, Jonathan Shear, Galen Strawson, eds, *Models of the Self* (Thorverton, UK: Imprint Academic, 2000), 1–24.

Stump, Eleonore and Kretzmann, Norman, 'Eternity', *The Journal of Philosophy*, Vol. 78, No. 8 (August 1981), 429–58.

Taylor, Richard, 'Spatial and Temporal Analogies and the Concept of Identity', in J. J. C. Smart, ed., *Problems of Space and Time* (New York: Macmillan, 1968), 381–96.

——, *Metaphysics*, third edition (Englewood Cliffs, New Jersey: Prentice-Hall, 1983), 41–50.

Tulving, Endel, 'Concepts of Memory', in Endel Tulving and Fergus Craik, eds, *The Oxford Handbook of Memory* (New York: Oxford University Press US, 2000), 33–43.

Tulving, Endel and Craik, Fergus, eds, The *Oxford Handbook of Memory* (New York: Oxford University Press US, 2000).

Whitehead, Alfred North, Chapter 3 'Time', *The Concept of Nature* (Cambridge: Cambridge University Press, 1982), 49–73.

Worms, Frédéric, *Introduction à Matière et Mémoire de Bergson* (Paris: Presses Universitaires de France, 1997).

——, '*Matter and Memory* on Mind and Body: Final Statements and New Perspectives', in John Mullarkey, ed., *The New Bergson* (Manchester and New York: Manchester University Press, 2000), 88–98.

Index